D1387593

FLORA

of

Staffordshire

FLORA
of
Staffordshire

FLOWERING PLANTS AND FERNS

by
E. S. EDEES

❁

DAVID & CHARLES : *NEWTON ABBOT*

0 7153 5563 5

*Set in Times by Avontype (Bristol) Limited
and printed in Great Britain by
Compton Press Limited, Compton Chamberlayne, Salisbury
for David & Charles (Publishers) Limited
South Devon House   Newton Abbot   Devon*

CONTENTS

LIST OF PLATES

Photographs not acknowledged above are by the author.

MAPS IN THE TEXT

PREFACE

The aim of this Flora is twofold. As no previous Flora of Stafford-shire has been published in book form and as the two that have appeared as appendices to journals (Bagnall 1901 and Ridge 1922–9) are brief and incomplete, it is necessary to make a fresh and thorough survey of the work of the early botanists. This necessity accounts for the historical bias of the present book. To express in a few words all that has been gleaned from the researches of the early botanists is impossible, but it is hoped that enough has been recorded to make it unnecessary for future workers, who are not botanical historians, to investigate all the early records again.

The second object of this Flora is to present a picture of the vegetation of Staffordshire as it was in the middle of the twentieth century both for comparison with past days and as a foundation for future work. Once again this can only be expressed in outline in a short book. But complete records, past and present, of every plant recorded for Staffordshire are preserved in manuscript.

This book has been written with an eye to a revised edition of the *Atlas of the British Flora* (Perring & Walters 1962). The 10km square of the national grid reference system was chosen as the recording unit in order that the records in the local Flora could be quickly and accurately transferred to the national *Atlas*. Moreover it so happens that the year 1930, adopted in the *Atlas* as the dividing point between old and recent, is also the most suitable date for separating the Staffordshire records in the same way.

But it was intended to go further than the *Atlas* and show by more detailed recording the frequency of species within each 10km square, at least as far as to say whether they were common, frequent or rare. Unfortunately, though the detailed recording has been done, it has not proved possible to print all the maps. However a complete range of maps is available for reference with the full manuscript in the author's possession.

The author's first duty is to provide a scientifically accurate record, but he hopes that the book will be readable and will rest not only on the shelves of botanical libraries, but also on the bedside tables of botanists! That is one reason for the uneven treatment of the species.

There is more to say about some plants than about others and, though some statements could be suppressed without scientific loss, the author has found so much joy in reading what the early botanists wrote that he quotes their words from time to time.

Finally this book is but a beginning. It is not a revision of an earlier Flora but a foundation on which to build. There is a great deal yet to be discovered in Staffordshire, critical work to be done on many groups of plants and an open field for ecological studies.

September 1971   E. S. Edees

THE VEGETATION OF STAFFORDSHIRE

DEFINITION OF STAFFORDSHIRE

Staffordshire is a midland county situated a little to the north of the centre of England. Its greatest length from north to south is about 56 miles (90km) and its extreme width from east to west nearly 38 miles (60km). Its area including Dudley, which, though part of the administrative county of Worcestershire, is enclosed by Stafford-shire and therefore botanically part of Staffordshire, is very roughly 750,000 acres.

The map of Staffordshire on which this Flora is based was drawn in 1962 and outlines the county as it was then. Since 1962 there have been adjustments to the boundary with Warwickshire in the south and south east, which we have had to ignore. But we cannot ignore areas of Warwickshire, Worcestershire and Shropshire which were part of Staffordshire in 1852, when H. C. Watson in the third volume of *Cybele Britannica* first designated Staffordshire vice-county 39. These comprise the eastern part of Sheriff Hales, a triangle in grid square SJ/7432, where the boundary used to be the stream flowing from east to west, most of Upper Arley and an urbanised area north of Birmingham. These deviations from the 1962 boundary are shown on the floral maps used in the present work by a broken line. To balance the losses there have been gains at Burton and Edingale and others, which are too small to be shown on our maps, at Tamworth, Warley Woods and Bobbington. Thus the map used in this Flora outlines Staffordshire both as it was in 1852, the botanical Stafford-shire which botanists continue to call vice-county 39, and as it was in 1962.

RELIEF

There are three well marked physical regions: the northern hills, the central plain and the southern plateau. In the north east there is an extensive area of moorland over 1,000ft high, rising to 1,684ft in Oliver Hill, the summit of the county. This is the southernmost part of the Pennine Chain. To the west of the moors, much of the land is between 400ft (122m) and 800ft (244m) high. The hill country is dissected by a series of parallel rivers which flow from north west to

3

south east directly into the Trent or into the Dove which joins the Trent on the Derbyshire border.

The central plain is a low lying tract of land watered by the river Trent, which rises on the moors near Biddulph and sweeps eastwards in a great curve.

The southern plateau protrudes like a wedge into the central plain and rises to 800ft at one point on Cannock Chase. On its western side the Penk flows north to join the Sow near Stafford and on the east the Tame flows north to merge with the Trent. Thus nearly the whole of Staffordshire is in the catchment area of the Trent. But in the extreme south west the rivers flow south to join the Severn.

GEOLOGY

The moorland country in the north east is composed of Carboniferous grits, shales and limestones. The rocks known as the Roches and Ramshaw Rocks, which form the roof of Staffordshire, are of Millstone grit and have been weathered into sharp edges. Beyond them to the east are the rounded hills of Morridge and further east still there is a compact area of limestone, through which the Hamps, Manifold and Dove have cut deep ravines. On the western side of the high hills lies the North Staffordshire coalfield and beyond that on the Cheshire and Shropshire boundary and along the southern edge of the coalfield eastwards to the limit of the county there is a border of Triassic sandstone.

In the south also much of the upland plateau is composed of Coal Measures and Triassic sandstone. The triangular South Staffordshire coalfield with its apex at Rugeley is bordered by a wide rim of sandstone. Cannock Chase consists of sandstone and Bunter pebble beds and there are large areas of the harder Keuper sandstone near Wolverhampton and round Lichfield. There is no Carboniferous limestone in the south, but there are small inliers of Silurian limestone at Dudley, Sedgley and Walsall.

For the rest the underlying rock is nearly everywhere Keuper marl. Most of the central plain consists of Keuper marl.

CLIMATE

There is a marked difference between the climate of the north and that of the south of Staffordshire. According to Myers (1945) the average annual rainfall for the south of the county is 25–30in and for the north 35–55in. At Leek and Mayfield in the north the wettest

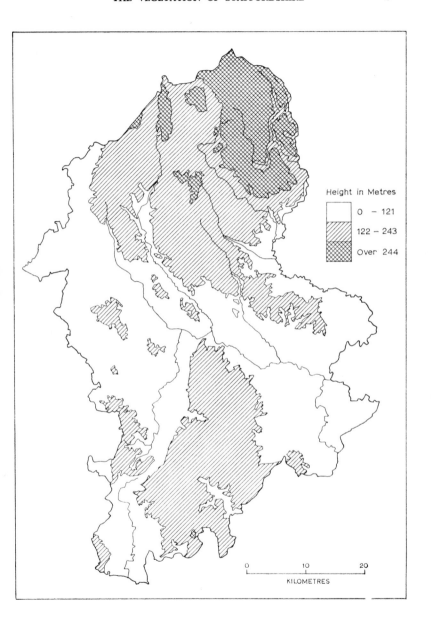

Relief Map

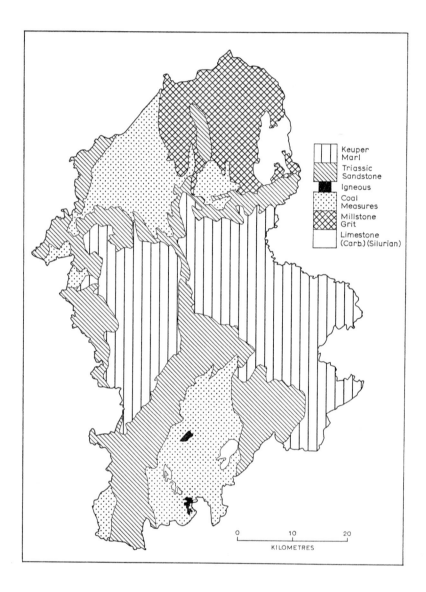

Solid Geology (simplified)

season is autumn and the wettest month December. In the south the wettest preiod is June to August and the wettest month August. Spring is the driest season over the whole county. It is much colder in the hills than in the lowlands and spring is often late in coming. At Buxton, which, though in Derbyshire, is close to the northern border of Staffordshire, there are on average 38 days with snow and 111 with ground frost every year.

POPULATION

The two main centers of dense population are the industrial areas based on the two great coalfields, the Potteries in the north and the Black Country in the south. During the nineteenth century the population of the county increased fivefold, from more than 240,000 in 1801 to nearly 1,240,000 in 1901, and today it is much greater. Nevertheless Staffordshire remains chiefly an agricultural county.

EFFECT OF LATITUDE

The geographical position of Staffordshire partly accounts for the diversity of its flora. It is the meeting place of north and south. For example *Valeriana, pyrenaica,* which is most at home in Scotland, and *Cardamine bulbifera,* which is almost confined to the south-eastern corner of England, are both found in Staffordshire. A glance at the *Atlas of the British Flora* (Perring & Walters 1962) shows that there are many species which are chiefly found south east of a line from the Humber to the Severn. Staffordshire adjoins this boundary on the north side and as a result we have several species which have a predominantly south-eastern distribution in the British Isles. These include *Bryonia dioica, Chenopodium polyspermum, Coronopus squamatus, Euphorbia exigua, Hordeum secalinum, Ononis spinosa, Pastinaca sativa, Ranunculus arvensis, Rorippa amphibia, Sison amomum, Thalictrum flavum* and *Veronica catenata.* Many are arable weeds and plants of low lying marshy ground. *Bryonia dioica* is one of the most interesting. It is abundant in hedgerows in the south of the county, but totally absent from the north, reproducing in the county the same clear cut pattern it has in the national *Atlas.*

Our link with the north is the Pennine Chain and most of the northern plants in the Staffordshire flora are upland species, though some are also found on Cannock Chase. The list includes *Crepis paludosa, Empetrum nigrum, Gymnocarpium dryopteris, Myrrhis odorata, Rubus chamaemorus, Salix pentandra, Thelypteris phegopteris,*

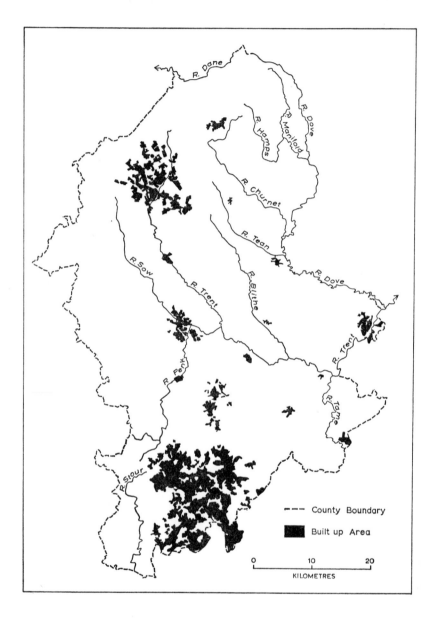

Population Map

Trollius europaeus and *Vaccinium vitis-idaea*. In addition to these we have related pairs of species with complementary geographical ranges, such as *Cirsium heterophyllum* and *C. dissectum*, *Dactylorhiza purpurella* and *D. praetermissa*, *Stellaria nemorum* and *Myosoton aquaticum*, where the first of each pair is a northern and the second a southern plant.

The meeting of east and west is less marked, though *Ceterach officinarum*, *Ulex gallii* and *Umbilicus rupestris* are three western species and *Hottonia palustris* an eastern one which extend into Staffordshire.

EFFECT OF ALTITUDE

Topographical factors are also important when we try to account for the distribution of plants within the county and the country. For example, *Scrophularia auriculata* has a southern distribution both in Britain and in Staffordshire perhaps chiefly because it is a lowland species. The same is true of *Ranunculus sceleratus*, which is an abundant plant in south Staffordshire but absent from the hills. On the other hand we have a few plants which are rarely found below 1,000ft. *Viola lutea* is a good example, though altitude is not the only factor restricting its distribution.

EFFECT OF GEOLOGY

The soils derived from the underlying rocks, unless deeply buried beneath alien deposits or too greatly modified by leaching, nurture plants which often reflect the geological structure of the county. The distribution map of *Primula vulgaris*, for example, marks out quite clearly the boundaries of the acid sandstone and the calcareous limestone and marl.

The attractiveness of the Staffordshire flora owes much to the Carboniferous limestone which links the county with Derbyshire. We have a long list of calcicolous species, including some which are confined to the Carboniferous limestone, at least in Staffordshire, and have a restricted distribution in Britain. The following are notable: *Daphne mezereum*, *Draba muralis*, *Galium sterneri*, *Hornungia petraea*, *Melica nutans*, *Polemonium caeruleum*, *Potentilla tabernaemontani*, *Polygonatum odoratum*, *Ribes alpinum*, *Saxifraga hypnoides*, *Silene nutans*, *Valerianella carinata*.

In the extreme south west of the county the sandy soil of Highgate Common and Kinver Edge bears a flora reminiscent of the Breckland

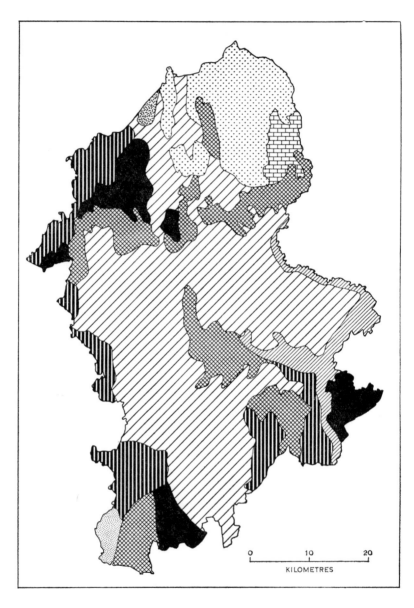

Soil Map (key on opposite page)

BASE CONTENT	GEOLOGY	TYPICAL SOIL ASSOCIATIONS	TYPICAL TEXTURE
	Glacial S. and G.	Podzols	Sand
LOW	Trias Sstn.	Podzols and Acid Brown	Sand and Sandy Loam
	Mill. Grit. Silica rock	Podzolized Acid Brown	Sandy Loam
	Mill. Grit. Shales	Peaty gleys	Clay
MEDIUM	Alluvium	Gleys	Clay or Silty clay
	Upp. Coal Meas. & Keuper Marl	Acid Brown	Sandy Loam
HIGH	Devonian	Leached Brown	Silt Loam
	Glacial till	Leached Brown and gleys	Loam
	Carb. Lstn.	Leached Brown (highly calc.)	Silt Loam

of East Anglia. Here we have *Cerastium arvense, Echium vulgare, Teesdalia nudicaulis, Viola canina* and several calcicolous species, such as *Myosotis ramosissima* and *Thymus drucei,* which link the south west with the north east.

Needwood Forest on the Keuper marl is famous for *Euphorbia amygdaloides,* which is found here plentifully in one of its most northerly British localities. Other characteristic species of what remains of this ancient woodland are *Carex strigosa, Daphne laureola* and *Tilia cordata.*

Salt beds near Stafford and Branston provide habitats for seaside plants, such as *Glaux maritima, Plantago maritima, Puccinellia maritima, Samolus valerandi, Scirpus maritimus, Spergularia marina* and *Triglochin maritima.*

In contrast to this wealth the sandstone and Bunter pebble beds, which give us Cannock Chase and Hanchurch Hills and much picturesque undulating scenery, have a poor flora. By keeping to the high ground it is possible to walk many miles over Cannock Chase and not record more than fifty species. But there are rich valleys. If the botanist starts from Sycamores Hill, which is the best place in

Staffordshire for him to begin the study of brambles, and walks down Oldacre Valley to Brocton, he may see *Drosera rotundifolia* and *Thelypteris palustris* and, if he knows where to look, *Carex dioica*, *Drosera anglica*, *Eleocharis quinqueflora*, *Parnassia palustris* and *Pinguicula vulgaris*, all rare plants in Staffordshire.

Lowland peat mosses used to be a special feature of the Staffordshire scene, but many of them have been drained. However, Chartley Moss, where *Andromeda polifolia*, *Drosera rotundifolia* and *Vaccinium oxycoccos* grow together in profusion, and Loynton Moss, which harbours *Carex elongata*, *Myrica gale* and *Ranunculus lingua*, have been preserved.

HUMAN INFLUENCE

When man first settled in what we now call Staffordshire most of the county was wooded and the rest marsh and moor. Gradually the trees were felled, the land was ploughed, houses were built, hedgerows planted, roads constructed. Then came the industrial revolution, the building of factory towns, the pollution of the rivers, the draining of mosses, the making of canals and railways, and in our day the mechanisation of farming, the spraying of crops and roadsides, the planting of coniferous forests and the obliteration of country lanes in the interests of wider and faster roads.

But it has not been all loss. New habitats have been created by the construction of reservoirs and the abandonment of worked out quarries, which plants have been quick to colonise. For example, soon after work ceased at Trentham gravel pit *Dactylorhiza purpurella* appeared in the wet sand.

It is worth remembering that some of the habitats man is now accused of destroying he was once blamed for creating. There was widespread objection in the early half of the nineteenth century to the making of railways and now we protest at their removal. It is the same with the canals. We have recently lost *Nymphoides peltata*, which with *Butomus umbellatus* and other water plants adorned the canal between Forton and Norbury. But we have not lost an original species of the Staffordshire flora, but rather one which man allowed to appear when he made the canal.

The presence of man and his works has sometimes brought about a redistribution of species. For example, before the industrial development of what we now call the Black Country *Paris quadrifolia* used to grow in the woods near Cradley. These were felled long ago and

Herb Paris disappeared from that part of the county. But it re-appeared round some of the marl pits which the farmers dug in their fields. Now the marl pits are being filled in and it will have to move on again! The marl pits were once very numerous in lowland Staffordshire, especially in the central plain, and, when they filled with water and developed a fringe of willow and alder, produced a rich and varied flora.

The building of towns and factories has obliterated much good country, but alien species quickly colonise waste ground and some of them, like *Senecio squalidus*, are beautiful in spite of their names. The general result of human interference with the countryside has been to reduce the quantity of many of the original native species but to increase the range of the flora.

Today there is a movement among countrymen to preserve the best of their heritage and already, thanks to the work of the National Trust, the Nature Conservancy, the Staffordshire Nature Conservation Trust and the generosity of many landowners, much of the Manifold Valley and Dovedale, a large part of Cannock Chase, Coombes Valley, the Downs Banks, Kinver Edge, Hawksmoor Wood, Chartley Moss, Loynton Moss, Black Firs and other botanically rich areas have been safe-guarded.

THE HISTORY OF FIELD BOTANY IN STAFFORDSHIRE

It is impossible in a few pages to do more than summarise the work of nearly 400 years. A detailed account of the period 1597–1839 has already been published (Edees 1948), but the full story of the following years awaits an author.

EARLY PERIOD (1597–1839)

The first plant to be recorded in print for Staffordshire, as far as we know, was *Vaccinium oxycoccos* (Gerarde 1597) and the second *Salix pentandra* (Johnson 1641). But very little is known about those early days. Celia Fiennes, who used to ride about England on horseback, came to Staffordshire in 1698 and stayed at Wolseley Hall for six weeks. She noticed the bilberries growing under the oak trees in the park, the bracken on Cannock Chase and the silver Trent, which meandered through meadows 'bedecked with hay almost ripe and with flowers'.

Other visitors were two university dons, John Ray from Cambridge and Robert Plot from Oxford. 'The excellent Mr Ray', as Gilbert White called him, did not find Staffordshire particularly interesting. In a short statement which he contributed to Gibson's edition of Camden's *Britannia* (1695) he says: 'The mountainous part of this country, called the Moorlands, produceth the same plants with the Peak-country of Derbyshire, the more depressed and level parts with Warwickshire.' He seems to imply that the botanist who knows Derbyshire and Warwickshire can afford to neglect Staffordshire. This is disappointing because Ray paid many visits to Middleton Hall, which is less than a mile from the Staffordshire border, and lived there for nearly three years (1672–5). There are, however, a few Staffordshire records in his *Catalogus Plantarum Angliae* (1670).

Plot came to Staffordshire at the invitation of Walter Chetwynd of Ingestre and published *The Natural History of Staffordshire* in 1686. Chapter six entitled 'Of Plants' contains little information of direct value to the scientific botanist. The common plants are all taken for granted. It is the strange and curious which attracts Plot, a foxglove of 'praeternatural whiteness' at Norton in the Moors, a white-fruited

14

elder near Combridge and a birch near Ranton Abbey with red leaves 'as if fresh blood had fallen on them'. He has a good deal to say about trees of exceptional size. An oak tree in Ellenhall park, which lay along the ground, was 'of so vast a bulk, that my man upon a horse of fifteen hands high, standing on one side of it, and I also on horseback on the other, could see no part of each other'. There was an apple tree at Leigh which could shelter eighty horsemen or 549 footmen from sun or rain. But perhaps his most delightful discovery (p 111) was that of a cow at Aston near Stone which had golden-tinted teeth. 'The ingenious Mr Lister, physician at York,' thought the cow had been feeding on the yellow mountain pansy.

Plot died in 1696 and thereafter for nearly a hundred years there is nothing further to report for Staffordshire. But towards the end of the eighteenth century there was an extraordinary flowering of interest in field botany within the county. Acland, Bagot, Bourne, Bree, Butt, Clifford, Dickenson, Forster, the Gisbornes, Pitt, Purton, Riley, Scott, Sneyd, Stokes, Wainwright, Waring, Withering, Wolseley, all these were midland botanists and many of them Staffordshire men who made their contribution to local knowledge between 1770 and 1820.

The greatest of them was William Withering, but the first to make an original contribution was Richard Waring, who lived in Cheshire. In 1770 he wrote a letter to Daines Barrington, well known as a correspondent of Gilbert White, which was published two years later in the *Philosophical Transactions*. There are references to fourteen Staffordshire species, most of which were seen at Willoughbridge or Maer in the north of the county.

William Withering (1741–99) was a Birmingham doctor whose career began in Stafford. As a botanist his strongest claim to remembrance is a book entitled *An Arrangement of British Plants*, of which (with variations of title) three editions were published in his lifetime and several more after his death. The second edition (1787) contains nearly 100 references to Staffordshire plants. His son tells us (1822) that his father's interest in botany began in Stafford when he used to collect plants for his future wife to draw. 'For her he explored the enamelled meadows watered by the Trent, the varied lawns of Shugborough, or the wild recesses of Haywood Park . . . and soon began to collect specimens for that herbarium which he afterwards rendered so complete.'

Meanwhile on the eastern side of the county two brothers, John and

Thomas Gisborne, were investigating the flora of the Weaver Hills and Needwood Forest. The Rev Thomas Gisborne (1758–1846) lived at Yoxall Lodge and became distinguished enough to win a place in the *Dictionary of National Biography*. The writer of his life there tells us that Needwood Forest meant as much to Thomas Gisborne as Selborne did to Gilbert White. In 1794 he published a book of poems called *Walks in a Forest*. He was an uncle of C. C. Babington, who became Professor of Botany at Cambridge and who often visited him at Yoxall. In a letter to Sir W. J. Hooker, dated 24 November 1834, Babington described him as 'an ardent botanist' who had lived more than forty years in Needwood Forest. When Gisborne died Babington went to Yoxall to look through his uncle's collection of plants and botanical books. He tells us (1897) that most of the plants were worthless from damp and were destroyed. Nevertheless there are twenty volumes of beautifully preserved specimens, representing about 600 species, in the herbarium of the British Museum. These were all collected by Gisborne at Yoxall Lodge or in its vicinity and most of them are dated 1791 or 1792. Among the books which came to Babington from his uncle and which are now at Cambridge is a copy of Hudson's *Flora Anglica* containing many manuscript records in Gisborne's hand.

John Gisborne (1770–1851) was also both a poet and a botanist. He lived for short periods at Wootton Hall, Holly Bush and Orgreave Hall. While he was at Wootton Hall (1792–5) he wrote a long descriptive poem called *The Vales of Wever*, which was published in 1797. There are references to several local plants in footnotes. But that is all that survives. Like Jean-Jacques Rousseau, who lived at Wootton Hall from 1766 to 1767, and Erasmus Darwin, who was botanising in the neighbourhood in 1766 when he met Rousseau, John Gisborne allowed most of his botanical knowledge of the district to die with him.

In 1796 William Pitt of Pendeford near Wolverhampton (1749–1823) issued a *General Survey of the Agriculture of the County of Stafford*. A preliminary report had been published two years earlier and there was a second edition in 1813. The preliminary report contains (pp 151–9) 'A List of Plants, Trees or Shrubs, natives of the County, remarkable for beauty or use, or for their medicinal, poisonous, or other singular qualities'. In the edition of 1796 this becomes (pp 210–26) a 'Botanical Catalogue' and is expanded from 100 items to more than 200. Besides this a list of 'Rare Plants in

Mr Sneyd's Woods and Walks' is increased from six items in the earlier book (pp 142-3) to twenty-seven in the later (pp 203-4), representing about forty-five species. The second edition of 1813 contains nothing new apart from a few footnotes.

The first volume of Stebbing Shaw's great work, *The History and Antiquities of Staffordshire*, 1798, contains (pp 97-115) a catalogue of native Staffordshire plants compiled by the Rev Samuel Dickenson of Blymhill (1730-1823). The list contains more than 800 names, of which about 650 are flowering plants and ferns. Most of these were recorded by the writer himself from the neighbourhood of Blymhill, but there are also numerous contributions from botanists in other parts of Staffordshire. In the second volume (pp 6-7), published in 1801, there are a few additional records contributed by Dickenson, Riley, Wainwright and Bourne. Dickenson was an excellent botanist and his habitat descriptions testify to his powers of observation and expression. His catalogue is the first comprehensive list of Staffordshire plants and a most worthy foundation work.

In 1805 D. Turner and L. W. Dillwyn published *The Botanist's Guide through England and Wales*. The section devoted to Staffordshire contains forty-one entries, but two are misplaced Oxfordshire records and most of the rest reproductions from earlier writers. A copy of *The Botanist's Guide* in the library of the National Museum of Wales contains many additional manuscript records, chiefly from the neighbourhood of Lichfield. The book once belonged to Dr J. A. Power (1810-86), whose herbarium is in the museum of the Holmesdale Natural History Society at Reigate, and notes and specimens often agree.

In 1817 Sir Thomas and Arthur Clifford of Tixall published *A Topographical and Historical Description of the Parish of Tixall in the County of Stafford*. This contains (pp 285-308) an attractive list of local plants under the title 'Flora Tixalliana: or a catalogue of the most remarkable phaenogamous plants to be found within a morning's ride of Tixall'. There are 268 entries, most of them initialled T.C. and many of them original. Included with the others are several from Stone contributed by Richard Forster. There is also a letter in the William Salt Library, Stafford, written by Forster in 1796, which contains a list of plants found near Stone. Forster was a surgeon at Stone, and Garner (1871) tells us that he had a pigtail and powdered his hair.

Other books published at this time may be briefly mentioned.

A Topographical History of Staffordshire by Wm. Pitt appeared in 1817. This contains (pp 101–21) a list of 680 Staffordshire flowering plants and ferns. In the same year the first volume of T. Purton's *A Midland Flora* was printed and the second volume four years later. In this work there are a few Staffordshire records by W. S. Rufford of Badsey, W. T. Bree of Allesley, W. Scott of Stourbridge and a particularly valuable set by a Mrs Acland of Lichfield. W. Scott published a book himself in 1832 called *Stourbridge and its Vicinity*, containing (pp 540–58) 'A Select Descriptive Botanical Catalogue' of 200–300 names, of which several are given Staffordshire localities. A much shorter list is contained in *A Descriptive and Historical Account of Dudley Castle* by L. Booker, 1825 (pp 107–9).

Babington's visits to Staffordshire are recorded in the *Memorials Journal and Botanical Correspondence of Charles Cardale Babington*, published by his widow in 1897. Nearly thirty Staffordshire plants are recorded in the *Journal*, all from Needwood Forest and the surrounding district. Some of these were communicated to H. C. Watson and published in *The New Botanist's Guide*, 1835–7, where in addition there are ten new records in the name of C. C. Babington or his cousin Churchill Babington. In the evening of his life Babington visited Yoxall again and we have the following entry in his journal for 30 September 1880: 'To Yoxall Lodge, to see the residence of my uncle Gisborne. Found it very much altered from what it was in his time; the house for the better, the grounds for the worse.'

Two papers appeared in *The Analyst* for 1837, one by Wm. Ick, entitled 'Remarkable Plants found growing in the Vicinity of Birmingham in the year 1836', and one by Miss M. A. Jackson with the title 'Catalogue of Some of the Rarer Species of Plants found in the Neighbourhood of Lichfield'. Finally, in 1839 J. Carter published in *The Magazine of Natural History* a document of first rate value for anyone studying the flora of the Churnet Valley. He called it 'A Few Observations on some of the Natural Objects in the Neighbourhood of Cheadle, Staffordshire'. In the course of the paper 63 plants are mentioned and at the end there is a list of a further 103.

MIDDLE PERIOD (1844–1929)

Robert Garner (1808–90) was the outstanding Staffordshire botanist of the nineteenth century. In 1844 he published *The Natural History of the County of Stafford*, which contains (pp 333–445) a list with localities of about 900 flowering plants and ferns. A short appreciation of this part of his book was issued by the present writer in 1944 and an account of Garner's life in 1950. Garner was a medical practitioner who lived in Stoke not far from the present railway station. He was one of the founders of the North Staffordshire Field Club in 1865 and contributed many papers to its *Transactions*.

One of Garner's friends was Samuel Carrington (died 1870), schoolmaster at Wetton and in later life locally famous as a geologist and archaeologist. In his early days Carrington was a keen botanist and used to draw and paint the wild flowers of his neighbourhood. There are 150 of these excellent reproductions in a small hand-made book now in my possession. The book is undated, but a memoir of Carrington written by Garner in 1870 suggests 1835 as an approximate date.

Another notebook made by G. Smith of Ockbrook and dated 26 May 1871 contains a list of 144 flowering plants and ferns found in or near Dovedale. Some of the records were contributed by Carrington, 'a very careful observer and recorder of local plants'. Painter (1889) tells us that Smith was vicar of Osmaston-by-Ashbourne from 1854–1871.

In 1853 Andrew Bloxam stayed for a few days at Warslow Hall. He spent a few hours in the Manifold Valley at Ecton and made a journey across the moors to Goldsitch Moss and Gradbach, but heavy rain curtailed his search for *Hymenophyllum wilsonii* which he had hoped to find.

O. Mosley's *The Natural History of Tutbury*, 1863, includes (pp 231–364) a comprehensive list of the flowering plants, ferns, mosses and fungi to be found within ten miles of Tutbury or Burton, compiled by Edwin Brown. Edwin Brown (1818–76) was one of a long line of botanists centred on Burton. Others were W. Birch, T. Gibbs, J. T. Harris, P. B. Mason, J. E. Nowers and J. G. Wells. These contributed to 'The Flora of Burton-on-Trent and Neighbourhood', which was published in four parts in the *Transactions of the Burton-on-Trent Natural History and Archaeological Society*, 1896–1901. Nowers and Wells (1890) also made an interesting study of the salt marsh vegetation at Branston. The herbarium of J. E. Nowers, which contains many plants from Needwood Forest and other parts of Staffordshire as well as from Burton, was given to the Darlington and Teesdale Naturalists' Field Club.

Other botanists published notes or lists of plants in the *Transactions of the North Staffordshire Field Club:* J. T. Arlidge, J. A. Audley,

S. Berrisford, J. Blaikie, C. Brett, D. Edwardes, R. W. Goodall, J. R. B. Masefield, J. H. Tylecote, J. Yates. The Rev D. Edwardes, who was headmaster of Denstone College, published a long list of plants found near Denstone in 1876 and Goodall (1882) tells us that in 1881 over 450 plants were found within seven miles of Denstone, including many ferns. But the ferns were becoming scarce. In 1880 he could find only one piece of *Asplenium viride* in Dovedale, until he met a hawker who had several in his basket and wanted to charge him eightpence for one. The basket also contained specimens of *Ceterach officinarum* and *Gymnocarpium robertianum* as well as commoner ferns, which had all been gathered in Dovedale on wet days when there were no people about.

J. R. B. Masefield (1850–1932), four times president of the North Staffordshire Field Club and a kinsman of the poet laureate, John Masefield, made a collection of plants from the vicinity of Cheadle in 1883, which is now in my herbarium. He was primarily an ornithologist, but he was a good all round naturalist. In 1909 he wrote a paper on Staffordshire ferns, attributing the decrease in the fern population to the collecting craze which he said reached its peak in 1870.

Samuel Berrisford (1859–1938) lived at Oakamoor, where he was employed as a blacksmith at the copper works. He made a valuable collection of plants from the Weaver Hills and Churnet Valley, totalling 700 species, which his daughter, Mrs J. Plant, gave to me. There are also nearly 400 local records pencilled in a copy of Anne Pratt's *The Flowering Plants of Great Britain*, which Mrs Plant allowed me to transcribe.

In the south of the county one of the most active botanists at this time was John Fraser of Wolverhampton (1820–1909). In 1864 he made a collection of 530 species of Staffordshire plants which was awarded the silver medal of the South Kensington Horticultural Society for the best county collection of the year. These plants with the rest of Fraser's extensive herbarium now belong to the University of Hull. There are more than a thousand sheets from Staffordshire, including many of great local value.

The north of the county was well served by clergymen, T. W. Daltry (1832–1904) at Madeley, W. H. Painter (1835–1910) at Biddulph and W. H. Purchas (1823–1903) at Alstonfield. Purchas and Painter were interested in critical genera, particularly *Rubus*, and won reputations outside the county as authors of Floras, Painter of

Derbyshire (1889) and Purchas (with A. Ley) of Herefordshire. Daltry inherited a love of botany from his father and passed it on to his son, so that we have through J. Daltry, T. W. Daltry and H. W. Daltry, whose lives were spent at Madeley, until H. W. Daltry left the county in 1951, more than a hundred years of unbroken local work. The plants they collected date back to 1840 and now form a valuable part of my herbarium.

Incorporated with a list of plants seen within five miles of Biddulph church, which Painter published in 1897, is another list from Stafford and King's Bromley compiled by Clifford Moore. Clifford Moore (1870–1944) was a chemist, who spent most of his life in Rugeley. His herbarium came to light in 1964 and was presented to the Rugeley and District Field Club (now the Mid-Staffs Field Club). There are specimens of about 400 species, including some garden plants, gathered mainly from the country between Stafford and Lichfield in 1889 and 1890.

Another botanist who must be mentioned to complete our survey of the nineteenth century was the Rev R. C. Douglas. In 1851 he supplied H. C. Watson with a list of plants found within three miles of Stafford for inclusion in *Topographical Botany*.

Further lists were made in the early years of the twentieth century. One of the best was compiled by the Rev T. Barns in 1912 of plants found within four miles of the vicarage at Hilderstone. But the foremost botanists of this period were J. E. Bagnall, H. P. Reader and W. T. B. Ridge.

J. E. Bagnall (1830–1918) was already well known as the author of *The Flora of Warwickshire*, 1891. In 1901 he published 'The Flora of Staffordshire' as a supplement to the *Journal of Botany*. This is a much slimmer work of seventy-four pages. It was intended to bring Garner's work up to date and is enriched by many new records supplied by the author and his contemporaries. In 1910 Bagnall wrote the botanical chapter for the *Victoria County History of Staffordshire*, dividing the county into five drainage areas, viz those of the Weaver, Dove, Trent, Sow and Severn and listing all the species recorded for each. This has an interesting introduction, but it is less important than the work of 1901. Bagnall's large herbarium, which contains many Staffordshire plants, is in Birmingham City Museum.

H. P. Reader (1850–1929), a Roman Catholic priest of the Dominican Order, was a first-rate botanist and made valuable contributions to the Floras of Leicester and Gloucester before coming to Hawkesyard in Staffordshire. He acquired an intimate knowledge of the flora of Rugeley and Cannock Chase and published papers about it in the the *Journal of Botany* and the *Transactions of the North Staffordshire Field Club*. His Staffordshire herbarium, now housed in the City of Stoke on Trent Museum, comprises about 800 beautifully mounted specimens gathered by himself between 1917 and 1923, together with a further 100–200 sheets collected by L. K. Clark, P. P. Thornton and F. D. Murray, his friends and colleagues.

Finally we must consider the work of W. T. Boydon Ridge (1872–1943), who spent his whole life in north Staffordshire. He was twice president of the North Staffordshire Field Club and chairman of its botanical section from 1904 to 1941. He taught botany at Hanley High School and led many field excursions in search of plants. His knowledge of the local flora led to his writing 'The Flora of North Staffordshire', which was issued in eight parts as appendices to the *Transactions* of the Field Club, 1922–9. In his early days Ridge made a small collection of dried plants, but the specimens were neglected and were later found to be badly ravaged by insects. The fifty or so it was possible to save are now in my herbarium.

RECENT PERIOD (1930–70)

The publication of Ridge's Flora marked the end of an epoch. The present writer came to Staffordshire in 1930, though it was not until 1941, when he succeeded Ridge as chairman of the botanical section of the North Staffordshire Field Club, that the gathering of records for a new Flora to embrace the whole county was seriously begun.

At first the civil parish was made the unit. Some of the smallest parishes were grouped together and the largest divided, making 224 areas to investigate. Garner had said that before the coming of the railways he had walked through every lane in the county in the cause of natural history. By 1941 most of the lanes had become country roads and the walker had trains and buses to take him to his starting point. In fourteen years the work was done and 224 lists of plants had been compiled.

Then in 1956 it was decided to start again and explore the county more thoroughly. Parish boundaries were abandoned and the squares of the national grid were used instead. It was thought that if twenty-five lists were made for every 10km square, then the 10km squares at any rate would have been explored with reasonable thoroughness. Accordingly each 10km square was divided into twenty-five smaller units, measuring 2 x 2 kilometres and representing about 1,000 acres of country. These were called minor squares and the 10km squares major squares. Once again the botanist set off on his travels, but now he had a motor car to take him to his starting point and even along some of Garner's original lanes. All the 800 minor squares, except for some in built-up areas, were visited three times, in the spring, in the early summer and in the late summer, and an average of about 200 species was recorded for each of them.

During the period 1930–70 the following have also recorded plants for Staffordshire and their help is most gratefully acknowledged:

Amphlett, Mrs	Deacon, Rev E.
Andrews, C. E.	Deakin, J.
Armitage, J.	De Nicolas, Mrs J. M.
Arnold, G. A.	Dodd, Mrs D. S.
Arnold, M. A.	Dony, Mrs J. G.
Baker, Mrs C. J. W.	Druce, Dr G. C.
Bates, Dr G. H.	Dutton, P. C.
Bates, G. L. D.	Edwards, J.
Beasley, F.	Elliot, Rev E. A.
Beaumont, Miss M. E.	Foskett, Miss D.
Bemrose, G. J. V.	Fowler, B. R.
Bennett, S. A.	Fox, Dr W. A. J.
Bentley, B.	Frost, Miss W.
Bigwood, Miss M. H.	Gibbons, Miss E. J.
Blizzard, Mrs A. E.	Goodway, Dr K. M.
Boniface, R. A.	Graddon, W. D.
Bowman, A. R. A.	Green, Mrs C.
Brenan, J. P. M.	Green, P. S.
Brown, R. H.	Griffiths, D. T.
Bryan, B.	Haigh, D. J. R.
Burges, Dr R. C. L.	Halden, A. J.
Burne, S. A. H.	Hall, F. T.
Butcher, Dr R. W.	Hall, R. H.
Cadbury, Miss D. A.	Hardaker, W. H.
Carey, Miss R.	Harris, R. G.
Castellan, Mrs M. L.	Haszard, Mrs D. R.
Chapple, J. F. G.	Hawkins, R.
Clarke, C.	Heslop-Harrison, Prof J. W.
Colclough, F. M.	Hibbert, Miss C. M.
Curtis, Sir Roger	Hine, G. S.
Daltry, H. W.	Hitchens, R. J.
Davies, A. R.	Hodgetts, Dr J. W.
Day, F. M.	Hollick, Miss K. M.

Horne, J. S.
Howitt, R. C. L.
Hudson, Miss M.
Jacobs, V.
Jennings, Mrs M. B.
John, Dr A. H.
Kingston, Miss M.
Landon, Miss G. L.
Leech, J.
Leonard, Miss W.
Lewis, D.
Lewis, Rev E. S.
Linley, Mrs G.
Lousley, J. E.
Lovenbury, G. A.
Masefield, J. R. B.
Matthews, Miss E.
Matthews, Miss J.
Meredith, W. D.
Meynell, Miss D.
Milne-Redhead, E.
Nelmes, E.
Orgill, Miss M. C.
Owen, H.
Parker, T. E. C.
Pendlebury, J. B.
Raybon
Richards, M.
Ridge, W. T. B.
Roscoe, Miss C.
Rutter, E. M.

Sell, P. D.
Shaw, H. K. A.
Slater, F. M.
Slater, Miss M.
Smith, M. E.
Smith, T.
Steele, H. J.
Stubbs, F.
Summerhayes, V. S.
Taylor, Sir George
Taylor, J. H.
Thomas, C.
Thompson, H. V.
Thompson, J.
Thompson, M. A.
Thornton, P. P.
Torrance, W. G.
Tutin, Prof T. G.
Wain, Miss D. E.
Walker, Miss K. B.
Wallace, T. J.
Warren, C. R.
Warren, R. G.
Watkin, A. E.
Watson, Miss P.
West, Dr C.
West, Mrs R.
Wilmott, A. J.
Wilson, Mrs L.
Woodhead, J. E.
Young, Dr D. P.

PRESENTATION OF THE RECORDS

To save space, adventives (unless they have occurred in several places), garden escapes (except for a few that have been frequently recorded or are long established) and species presumed to have been incorrectly recorded, have been omitted. Most of the omitted aliens come from Burton and are listed by Curtis (1930) and Burges (1944). The presumed errors include plants like *Galium boreale* and *Juncus biglumis*, which are most unlikely to have occurred wild in Staffordshire. Nevertheless when botanists find plants in Staffordshire, which are not included in this Flora, they should not publish them as new county records until they have been checked against the author's full manuscript list of recorded names.

The sequence of species and Latin nomenclature are that of Dandy (1958 and 1969), with a few exceptions in the genus *Rubus*, while the English names are taken by kind permission of Dr J. G. Dony from the 'Recommended List of English Names of Wild Plants' prepared by a working party of the Botanical Society of the British Isles.

The words used to describe the distribution of the species carry their everyday meaning. It is assumed that the reader will know what the author means when he says that one species is plentiful and another infrequent. For precise information he is directed to the statistical summary which follows the general description. This is often clear enough by itself without the addition of explanatory words. For example, the formula A–z, 796 should be enough to show that *Ranunculus repens* is common throughout the county.

Records for other than rare species are sometimes quoted for their scientific or literary interest, particularly if they are old. First records are also given, by name and date where the reference is unlocalised, but otherwise in full or abbreviated in such a way that nothing important is lost. These are the earliest records known to the author, but further research may antedate some of them.

The statistical summary lists the 10km squares of the national grid for which the species has been recorded. For convenience of quotation these squares are represented by letters as follows:

25

A	=	SJ/86	N	=	SK/04	a	=	SK/12	n	= SK/10
B	=	SJ/96	P	=	SK/14	b	=	SK/22	p	= SK/20
C	=	SK/06	Q	=	SJ/63	c	=	SJ/71	q	= SO/79
D	=	SK/16	R	=	SJ/73	d	=	SJ/81	r	= SO/89
E	=	SJ/75	S	=	SJ/83	e	=	SJ/91	s	= SO/99
F	=	SJ/85	T	=	SJ/93	f	=	SK/01	t	= SP/09
G	=	SJ/95	U	=	SK/03	g	=	SK/11	u	= SP/19
H	=	SK/05	V	=	SK/13	h	=	SK/21	v	= SO/78
J	=	SK/15	W	=	SJ/72	i	=	SJ/70	w	= SO/88
K	=	SJ/74	X	=	SJ/82	j	=	SJ/80	x	= SO/98
L	=	SJ/84	Y	=	SJ/92	k	=	SJ/90	y	= SP/08
M	=	SJ/94	Z	=	SK/02	m	=	SK/00	z	= SO/77

Three of these squares, the squares lettered i, q, u, include such small areas of Staffordshire that they are omitted from the summaries of widely distributed plants. Thus A–z indicates an occurrence in every square from A to z with the possible exception of i, q, u. Bracketed letters are used for pre-1930 records which have not been confirmed. The others are the post-1929 records of the present survey, which are referred to as recent records.

It is impossible in a short space to indicate the number of minor square records for each major square, but the grand total out of a maximum of 800 is given at the end. These are recent records and the majority of them date from 1956. The statistical summary is omitted for rare species when all the records are quoted in full.

The reference system has been made as simple as possible. Recorders' names are generally omitted, except for first records, unless needed to indicate a herbarium specimen or a book or article listed in the bibliography. A date followed by an exclamation mark refers to a herbarium specimen and if no name is given the collector is the author or one of his correspondents. Where there is no exclamation mark and no name either with the date or in the text, the reference is to the botanical report for the year in the *Transactions of the North Staffordshire Field Club* or its successor the *North Staffordshire Journal of Field Studies*.

INTERPRETATION OF THE MAPS

The large squares are the 10km squares of the national grid, conveniently called major squares, and the smaller squares blocks of four km squares, which the author likes to call minor squares, but which nowadays are often termed tetrads. Solid circles are used for recent, that is post–1929, records and hollow circles for unconfirmed earlier records which can be accurately plotted.

At least 95 per cent of the solid circles represent the personal observations of the author. This does not mean, of course, that the author's contribution to the Flora is nineteen times greater than that of all other contributors put together, but simply that common plants, which do not get recorded in print, are much more numerous than rare plants which do.

Perhaps it is also important to say that, though no square has been exhaustively explored, every square has been explored more or less equally. One of the many disadvantages of trying to survey a whole county single-handed is that fewer plants are recorded per tetrad. But there is a compensating advantage. When the task is finished and the maps drawn, the distribution patterns may be more reliable, because the squares have been evenly explored.

The floral maps published in this book have been selected to show the patterns of distribution most clearly discernible in Staffordshire. Some of them are difficult to interpret, but the topographical, geological, soil and population maps should help to explain them.

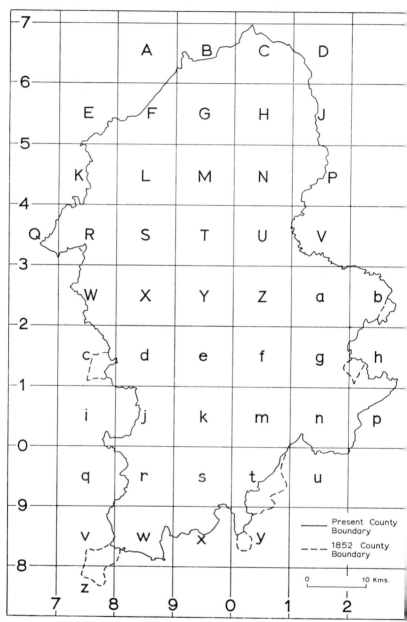

The Major Squares with Lettered References

PTERIDOPHYTA
LYCOPSIDA
LYCOPODIACEAE

Lycopodium selago L.   Fir Clubmoss
Unconfirmed records for Needwood Forest (Withering 1801); Barr Common (one plant in 1836), Maer Heath, Swynnerton Heath (Newman 1843); Offley Hay (Garner 1844); weed in garden at Bar Hill, Madeley (BEC 1920 Rep); wood near Cheadle (Ridge in Flora).

L. inundatum L.   Marsh Clubmoss
Unconfirmed records for Norton Bog (Withering 1801); Dimminsdale (Carter 1839); Offley Hay (Garner 1844).

L. clavatum L.   Stagshorn Clubmoss
Dry heaths in central and western Staffordshire: rare and decreasing: (F,) L, (M–N), R–S, (U), e, (m), (r), (t), (w), 6. There are old records for twenty places, including one for Axe Edge (Garner 1844), which may belong to Derbyshire, but which seems to justify Ray's statement (1670), 'On the mountains of Staffordshire.' Garner (1844) said it was sometimes seen in village fire-places as an ornament during summer. It was seen in Trentham Park in 1866 by the North Staffordshire Field Club on one of their first excursions and persisted there until recent years. In 1951 three or four small plants were found in the heather on Tittensor Chase and in 1958 it was noticed in Bishop's Wood. It survives on Cannock Chase where it was first discovered by Withering in 1787. A fine colony was reported in 1966 growing under a birch tree in a disused gravel pit at Brocton. In 1969 a dozen very small plants were found in a young pine plantation on Maer Hills.

SPHENOPSIDA
EQUISETACEAE

Equisetum hyemale L.   Rough Horsetail
Unconfirmed records for Prestwood Farm near Wednesfield (Pitt 1796); Lichfield (Jackson 1837); Rolleston Hall (Brown 1863).

E. fluviatile L.   Water Horsetail

'In shallow pools, common' (Dickenson 1798): B–P, R–T, W–n,
r–t, w, y, 204.

E. palustre L.   Marsh Horsetai

Marshy places in fields and woods and on the moors: A–g, j–n, r,
t–y, 224. Pitt, 1796.

E. sylvaticum L.   Wood Horsetail

Locally abundant in the north, growing in damp woods and by
streams in the gritstone cloughs, but rare in the south: B–D, F–H,
K–P, R, T, (Z–a), d–f, m, u, (z), 104. 'In the dingle at Cotton Hall'
(Dickenson 1798). Garner (1871) was right to say that it is seldom
absent from coal shale, at any rate in the north. (Map p 196).

E. arvense L.   Common Horsetail

Cultivated fields, roadsides, railway banks and waste land: A–z,
773. Pitt, 1794.

E. telmateia Ehrh.   Great Horsetail

Marshy woods and other wet shady places: frequent: B, F, H,
K–L, N–U, W, Y–a, (b), e, (f), g, j, (r), s–t, w, 32. Dickenson, 1798.
(Map p 196).

PTEROPSIDA
OSMUNDACEAE

Osmunda regalis L.   Royal Fern

Unconfirmed records for Aqualate Mere (Dickenson 1798);
Chartley Moss (Carter 1839); Willoughbridge (Garner 1844);
Cranmoor Wood (Fraser 1888!); marsh near Lichfield race course
(Power in Bagnall 1901). Recent records for the moors west of the
Royal Cottage, where there is a thriving plant growing by a stream
at 1,150ft (1955); Balterley Heath (1961!), where it was first
recorded by Pinder (Newman 1843); Weeford (1964). The ferns on
Chartley Moss were 'ruthlessly' collected and had become rare by
1863. Twenty years later they were considered to be extinct (Tylecote
1886).

HYMENOPHYLLACEAE

Hymenophyllum tunbrigense (L.) Sm.   Tunbridge Filmy-fern

Discovered by G. Mountney in 1907 at the northern end of Wootton
Park. P. Harvey, who had been shown the fern in the 1930s, told me

that there was a considerable quantity of it underneath a rock overhanging the valley and facing west, SK/095453. But he and I searched for it in vain in 1958. A specimen in Masefield's herbarium, dated 1909, was confirmed by J. A. Crabbe in 1963.

H. wilsonii Hook.   Wilson's Filmy-fern
One unconfirmed record for Gradbach, where the fern was said to grow in clefts of the rocks (Hewgill in Garner 1844). In 1853 Bloxam looked for it at Gradbach bridge but did not find it.

DENNSTAEDTIACEAE

Pteridium aquilinum (L.) Kuhn   Bracken
Heaths and birch woods: A–U, W–z, 655. 'In Kankwood' (Celia Fiennes 1698).

BLECHNACEAE

Blechnum spicant (L.) Roth   Hard Fern
Upland moors: B–D, F–H, K, M–P, R, T–U, (Y), (a), (c), e, (f), (k), (y), z, 79. 'White Sitch Pool, on the dam' (Dickenson 1798). (Map p 196).

ASPLENIACEAE

Phyllitis scolopendrium (L.) Newm.   Hartstongue
Rocky woods, shaded cliffs, damp walls, well mouths and disused pit shafts: native on the Carboniferous limestone, but elsewhere less common and often doubtfully native: B, H–L, N–S, W, Y–Z, (a–b), d, (f), (n), p, r, t, v, (w), y, 35. 'Woods at Belmont' (Sneyd in Pitt 1794). In 1909 Masefield said that it had disappeared from many places near Stone where it had once been abundant. Today, except on the limestone, it is unusual to find more than a few plants in any one place. It still grows at Heleigh Castle, where it was recorded in 1843 (Newman), and there are fine examples in a steep wood near Barrowhill (Rocester) with fronds two feet long (1968).

Asplenium adiantum-nigrum L.   Black Spleenwort
Rocks, old walls and mine shafts: H–J, (K), (N), Q–T, c, (e), h, n, w, 10. Ashwood Wells (Scott 1832).

A. trichomanes L.   Maidenhair Spleenwort
Walls and crevices of rocks, especially on the limestone, where it is one of the commonest ferns: B, (C), H–J, (K), M–P, S, Y, d, (e–f), (n), w, 28. 'In clefts of rocks on Weaver Hill' (Dickenson 1798).

A. viride Huds.　　　　　　　　　　　　　　Green Spleenwort
Dovedale (Garner 1844), but no recent confirmation. Smith (1871)
said it grew on north facing rocks.

A. ruta-muraria L.　　　　　　　　　　　　　　　　Wall-rue
Rocks and walls: frequent throughout the county, but most
common on the limestone: (C), H–P, R–T, W–d, (e–f), g, i–k, n–r,
(t), vaw, 68. Weaver Hills (Dickenson 1798). (Map p 196).

Ceterach officinarum DC.　　　　　　　　　　　　Rustyback
Limestone rocks and lime-mortared walls: rare. Wetton, Beresford,
Beeston Tor, Dovedale (Garner 1844); walls of canal bridge
near Prestwood House, Enville (Bagnall 1897!); Waterhouses
(Berrisford 1902!); wall near Stone (Masefield 1909); plentiful on
a wall in Seisdon village (1946); one good plant on a rock in
Wolfscote Dale (1949); wall at Sutton (1954); Sunny Bank (1955);
Hall Dale (1955).

ATHYRIACEAE

Athyrium filix-femina (L.) Roth　　　　　　　　　　Lady-fern
By streams both in sheltered places and on open hillsides: abundant
in the north: B–a, c–g, i–z, 310. Dickenson, 1798. (Map p 197).

Cystopteris fragilis (L.) Bernh.　　　　　　Brittle Bladder-fern
Walls and rock crevices: common on the Carboniferous limestone:
(C), (F–G), H–J, (L), (N–P), 15. 'On a wall between Oakamoor
and Cotton Hall' (Dickenson 1798).

ASPIDIACEAE

Dryopteris filix-mas (L.) Schott　　　　　　　　　Male-fern
Woods and hedgebanks: A–z, 662. Dickenson, 1798.

D. pseudomas (Wollaston) Holub & Pouzar　　Scaly Male-fern
Woods: frequent in the north: B–C, G–L, N, S, W, (Y–a), 11.
Bagnall, 1901, as *Lastrea filix-mas* var *paleacea*.

D. cristata (L.) A. Gray　　　　　　　　Crested Buckler-fern
Fraser collected specimens from Kingston Pool near Stafford in
1885. The fern grew on the eastern side of the pool, but it seems to
have disappeared soon after it was discoverd. Now the pool has
disappeared too.

D. carthusiana (Vill.) H. P. Fuchs　　　　Narrow Buckler-fern
Swampy woods and thickets bordering peat mosses: local: E, G,
K–L, N, R–T, W–a, c, e–f, (g), n, t, z, 39. Carter, 1839. (Map p 197).

D. dilatata (Hoffm.) A. Gray   Broad Buckler-fern
Woods and hedgerows: A–U, W–z, 570. Carter, 1839.

Polystichum setiferum (Forsk.) Woynar   Soft Shield-fern
Rare or local in some of our richer woods: J, (K), (N), P, (T), (a),
(j), v, y, 6. Near Lichfield (Jackson 1837). Recent records for
Dydon Wood, above the waterfall (1949!); Arley woods (1959!);
Cheshire Wood (Shimwell 1968); deep drumble on north side of
Barrowhill near Rocester (1968). It should be looked for in
Needwood Forest (Marchington Cliff and Forest Banks) and in
Star Wood, Oakamoor, for which there are old records.

P. aculeatum (L.) Roth   Hard Shield-fern
Woods, especially in rocky places by streams, and shady roadside
banks: most common on the limestone but not confined to it:
(B), H–K, (N), P–R, T, W, a, (b), d, (g), (m–n), v, (x), (z), 24.
'Russell's Hall and Rowley' (Wainwright in Shaw 1801). Druce
recorded *P. aculeatum* x *setiferum* for Throwley Moor (BEC 1917
Rep).

THELYPTERIDACEAE
Thelypteris limbosperma (All.) H. P. Fuchs   Lemon-scented Fern
Common among the hills, especially in the gritstone cloughs:
infrequent elsewhere on acid heaths and in woods: B–C, F–H,
K–P, T–U, (W), Z, (a), e, (f–g), m, (n), (v–w), (z), 80. Near Lichfield
(Jackson 1837). (Map p 197).

T. palustris Schott   Marsh Fern
Swampy thickets: very local. Offley Hay (Garner 1844); Cannock
Chase and Chartley Moss (Brown 1863); Rolleston (Masefield
1909); Betley Mere (1951!); Balterley Heath (1961!). The records
for Offley Hay and Rolleston have not been confirmed, but the fern
persists on Chartley Moss and there is a vigorous colony in Oldacre
Valley on Cannock Chase (1951!).

T. phegopteris (L.) Slosson   Beech Fern
Ridge Hill and Madeley Manor (Pinder in Newman 1843);
Brownhills (Brown 1863); woods near Ludchurch (1869); Stanton
Wood (Smith 1871); Needwood Forest (Nowers 1881!); Wootton
(1891); Wetton (Ridge 1892!); Bishop's Wood (1897); Consall
(Berrisford 1902!); Cheadle (Masefield 1909). In 1909 Masefield
said that, though rare, this fern could still be found in several
ravines in the north of the county. Today it is known only for

Burnt Wood, where it grows in a fir plantation by the side of a stream (1951!).

Gymnocarpium dryopteris (L.) Newm.   Oak Fern
Unconfirmed records for Needwood Forest (Pitt 1794) and for several woods in the Churnet and Dane valleys. It was often seen in Trentham Park, where Garner (1857) tells us it grew in Spring Valley. Today it is known only for Dydon Wood, growing profusely on a shaded rock above a waterfall (1949!), and Ludchurch (1949).

G. robertianum (Hoffm.) Newm.   Limestone Fern
Limestone screes and rock crevices: rare. 'Dovedale, Staffordshire' (Smith 1871); Caldonlow, in a disused quarry (1962); Rushton, station platform (1965!); near Drabber Tor (1969).

POLYPODIACEAE

Polypodium vulgare L. sensu lato   Polypody
'On old walls, shady places, and at the roots of trees' (Pitt 1796): most of the records are for the limestone in the north east and the sandstone in the west: B–C, H–L, N–S, U, W–X, Z, c–d, f, k, r, v, (y), 88. The segregates *P. vulgare* L. sensu stricto and *P. interjectum* Shivas occur in Staffordshire. Of ten specimens submitted to J. A. Crabbe in 1971 two were named *P. interjectum:* Oakamoor (Berrisford 1902!); Whitmore, SJ/832415 (1967!). (Map p 197).

MARSILEACEAE

Pilularia globulifera L.   Pillwort
Unconfirmed records for Hatherton and Offley Hay (Pinder in Garner 1844).

AZOLLACEAE

Azolla filiculoides Lam.   Water Fern
In the river Sow near Shugborough, F. Beasley (1968).

OPHIOGLOSSACEAE

Botrychium lunaria (L.) Sw.   Moonwort
Hill pastures, gritstone edges and sandy commons: rare and irregular in appearing: C, (F–G), H–J, (K), (M), N, (P), (U), (f), w, 9. 'On coal pit banks near Stourbridge' (Hill in Withering 1787). In 1839 Carter reported seeing hundreds of plants in a pasture near Wootton in the space of a few yards. In 1877 T. W. Daltry gathered

a specimen on Camp Hills, Maer, which is nine inches long.
Usually the plants are much smaller and much rarer. Recent
records for Blore (1932); Hollinsclough (1947); Back of Ecton, 'by
the wall of the old mine above this cottage' (Miss Roscoe in 1948!);
Highgate Common (1955); Coombes Valley (1955); Caldonlow
quarries (1969); Flash; several places in the Manifold Valley.

Ophioglossum vulgatum L.   Adderstongue
Damp grassy places: frequent: B, (F–G), H–J, (K), L, (M–N), P,
(T), X, (Y), a, (b), d, (f), g, (n), r, t, (w–x), 17. Near Blymhill
(Dickenson in Withering 1787).

SPERMATOPHYTA
GYMNOSPERMAE
PINACEAE
Pinus sylvestris L.   Scots Pine
Introduced, but regenerates freely in gravelly and peaty soils:
B–g, j–r, t–w, 183. 'Frequent; flourishes well in bogs' (Garner
1844).

TAXACEAE
Taxus baccata L.   Yew
Native in the Manifold Valley and Dovedale on the limestone
cliffs and perhaps elsewhere in woods and fields, but most of the
yews which remain in our fields, like those which occur in the
churchyards, were probably planted: B–C, E–P, R–w, (y), z, 221.
Plot, 1686.

ANGIOSPERMAE: DICOTYLEDONES
RANUNCULACEAE
Caltha palustris L.   Marsh-marigold
Wet woods and marshy fields, seen to perfection by rills in moor-
land pastures: A–w, y, 438. Near Yoxall Lodge (Gisborne 1792!).

Trollius europaeus L.   Globeflower
Hayfield at Gib Torr (1947) and a few plants in Brand Plantation
(1950). 'Woods at Belmont' (Dickenson 1798). When I visited the
hayfield at Gib Torr on 7 June 1950 there were about 1,000 blooms
intermingled with *Caltha palustris* and *Geum rivale*. Since then the

field has been ploughed and drained, but on 29 May 1965 there were 200 blooms in a wet patch that had been spared.

Helleborus foetidus L. Stinking Hellebore
Belmont (Sneyd in Pitt 1794); 'Old ditch bank in Tixall Ley Park, under the thorns near the Crabtree stocks' (Clifford 1817); near Leek (1907); adventive near Walton (BEC 1931 Rep).

H. viridis L. Green Hellebore
Here and there in thickets, orchards and field corners as a relic of cultivation: (F), H, (J), (N), P, a, (b), f, 5. Garner, 1844. Recent records for Yoxall Lodge (1945); wood behind Okeover Hall garden (1951); Foxholes, Hanbury (1957); Longdon, under bushes round a pit, SK/064137 (1962!); Bradnop, field along lane to Morridge Side (1967!).

Aconitum napellus L. Monkshood
A garden plant which is sometimes found semi-wild on river banks and in thickets near houses. First recorded by Forster (1796) near Coppice Farm, Stone: 'Whether this plant may properly be considered as indigenous or has been formerly an outcast from some garden I cannot say, but it has now established itself on the banks of the water and I do not recollect to have seen it in any of the gardens near the spot.' Garner (1844) thought it 'really wild' on the banks of the Churnet two miles below Cheddleton.

Anemone nemorosa L. Wood Anemone
Deciduous woodland and sometimes on hedgebanks and in fields bordering woods: A–g, j–n, r–w, y–z, 247. Abundant in Needwood Forest on the Keuper marl and in the wooded parts of the Manifold Valley on the Carboniferous limestone. It is also a frequent constituent of upland hayfields. In the upper part of the Hamps Valley, between Mixon and Feltysitch, there are colonies growing in open pasture at 1,200ft, where the only trees are a few scattered hawthorns. The plants are most numerous on the land sloping to the river. In the south of the county Reader (1923) found it abundant in meadows by the Trent and the earliest record (Stokes in Withering 1787) is for pastures in the neighbourhood of Stourbridge, 'some of which are almost white with it when in blossom.' The flowers vary much in colour, 'now robust in virgin white, now blushing with faint crimson' (Gisborne 1794). On the limestone in the Manifold Valley there are colonies of plants which combine deep reddish-purple flowers with narrow 'petals'. (Map p 198).

Clematis vitalba L.   Traveller's-joy
Naturalised in a few places in hedges near gardens: (J), Q, X–Y,
(a), b, d, f, j, n, r–s, w, (z), 13. Scott, 1832.

Ranunculus acris L.   Meadow Buttercup
Meadows: A–z, 760. Pitt, 1794.

R. repens L.   Creeping Buttercup
Wet meadows, a troublesome weed in clay soils: A–z, 796.
Dickenson, 1798.

R. bulbosus L.   Bulbous Buttercup
Common in the spring in dry pastures: B–z, 490. Dickenson, 1798.

R. arvensis L.   Corn Buttercup
Cornfields and as a casual on waste ground, once common, now
rare: C, L, (M–N), (S), X, Z, (a–b), (f–g), j, p, (r), s–t, v, 11. 'Cornfield near Hadley End' (Gisborne 1792!). 'I have seen this plant so
abundant as to be very injurious to a wheat crop on strong moist
land' (Pitt 1794).

R. sardous Crantz   Hairy Buttercup
Recorded for arable fields in the 19th century: (L), (N), (W), (Y),
(b), (f), (r). Garner (1844) said it was abundant in clayey fields
about Stafford and High Offley and Fraser (1864) found it in
several places on the west side of Wolverhampton. The earliest
record is for the neighbourhood of Newcastle (Pitt 1817) and the
latest for Bentley Farm near Rugeley (Reader 1923).

R. parviflorus L.   Small-flowered Buttercup
Sandy fields: rare: (T), (b), (d), j, (r), (w–x), 2. Blymhill (Dickenson
1798). Recorded recently for Brewood, where it was found in great
abundance and luxuriance in the neglected gardens of two tumble-
down cottages, one, Dingle Cottage, near Boscobel House and the
other by the river Penk north east of the village (1961!).

R. auricomus L.   Goldilocks Buttercup
Woods and shady hedgerows, especially on the limestone and
marl: common on the east side of Staffordshire and at Arley, but
infrequent elsewhere: H–J, (K), L, N–P, S–W, Y–b, d, (e), f–h,
n–p, (r), (t), v, z, 69. Dickenson, 1798. (Map p 198).

R. lingua L.   Greater Spearwort
In reed beds on the swampy margins of lakes and pools: rare:

(G), (K), (N), U, W, (X–Y), Z, (a), d, (e), (m), t, y, 6. Kingston Pool near Stafford (Stokes in Withering 1787). At the present day it is known for two pools near Uttoxeter, for two near West Bromwich, for the lake in Weston Park and for Loynton Moss.

R. flammula L. Lesser Spearwort
Marshy places: particularly common in wet peat by moorland streams: A–U, W–a, c–g, j–t, w, y–z, 330. Pitt, 1796. (Map p 198).

R. sceleratus Celery-leaved Buttercup
Ditches and the muddy borders of shallow pools: common in lowland Staffordshire, but absent from the hills: E, K–M, Q–u, w–y, 317. Dickenson, 1798. (Map p 198).

R. hederaceus L. Ivy-leaved Crowfoot
Ditches, rivulets, pond margins, in mud and shallow water: B–C, F–H, K–P, R–U, W–a, (b), d–f, m–n, (r), t–u, y, 69. Nearly twice as common as *R. omiophyllus* over the whole county, but less common among the hills. Brown (1863) seems to have been the first local botanist to distinguish the two species. (Map p 199).

R. omiophyllus Ten. Round-leaved Crowfoot
Wet muddy places: local or rare in the south, but frequent in the north: B–D, F–H, M–N, R, T–U, (W), a, (b), d–f, (m), n, (t), 40. Chorlton Moss (Daltry 1840!). (Map p 199).

R. aquatilis L. sensu lato Water Crowfoot
The water crowfoots as a group are common throughout Stafford-shire in ponds and rivers, but it is not possible to classify the old records in the absence of specimens. The following brief account rests almost entirely on the thirty-five local specimens in my herbarium which Dr C. D. K. Cook determined for me in 1968. The specimens of Bagnall, Fraser, Nowers and Reader have not been examined by a modern authority. The aggregate species was first recorded for Staffordshire by Ray (1670): 'In the river Tame about Tamworth.' This was one of the 'fennel-leaved' species (Withering 1787). Dickenson (1798) wrote: 'The variety of leaves in this plant is truly admirable. Those which are under water are capillary, or finely cut, like fennel leaves. Those on the surface are roundish, five lobed and notched. Its numerous beautiful white flowers are a great ornament to our ponds and lakes in May and continue till July.'

R. fluitans Lam.   River Water-crowfoot
In strongly flowing water: in the Blithe at Blithfield (1948!), where
the reservoir now is, and at Nethertown (1956!); Lonco Brook at
Whitleyford Bridge (1954!); river Penk at Acton Mill Bridge (1968).

R. circinatus Sibth.   Fan-leaved Water-crowfoot
Canals and ponds: Alton (Berrisford 1902!); Lichfield (1948!);
Hanchurch Pools (1949!); Stafford, dyke at St Thomas (1956!).

R. trichophyllus Chaix   Thread-leaved Water-crowfoot
Allimore Green Common (1968!), specimen not seen by Cook.

R. aquatilis L. sensu stricto
Ponds: Alton (Berrisford 1902!); Stone, in a pool by the roadside
opposite Pool House (1948!); West Bromwich, runnel beside marl
pit (1949!); Blymhill, pit in a field south of Aquamoor (1954!);
Abbots Bromley, pond opposite Seedcroft near Mill Green (1956!);
Branston, in one of the lily pits (1956!); Hamstall Ridware (1956!).

R. peltatus Schrank   Common Water-crowfoot
The commonest species of ponds and still water: B, F–G, K–N, S,
W–Y, a, c–e, 19. Madeley Manor moat (Daltry 1840!).

R. penicillatus (Dumort.) Bab.
Flowing water: Ellastone, in the deep flowing feeder between the
river Dove and the lake in Calwich Park (1947!); in the Dove near
Alstonfield (1953!), as var *calcareus* (Butcher) C. D. K. Cook.

R. ficaria L.   Lesser Celandine
Woods, meadows, hedge bottoms and damp sheltered places:
A–z, 577. Near Yoxall Lodge (Gisborne 1792!). Subsp *bulbifer*
(Marsden-Jones) Lawalrée has been recorded for a wood north
west of Hay Head Farm near Walsall (1962!) and for Hopwas Wood
near Tamworth (1962!) and is probably common.

Adonis annua L.   Pheasant's-eye
A rare casual of cornfields and waste ground. Single plants were
found in a cornfield near Swynnerton in 1930 and on waste ground
at Nelson Hall in 1947. There are earlier records for Tutbury
(Brown 1863); Burton (Nowers 1895!); Oakamoor (Berrisford
1902!); near Lichfield (1914).

Myosurus minimus L.   Mousetail
Recorded more than 100 years ago for cart ruts in low-lying arable

fields on the east side of the county. 'In a cornfield near Elford, 1800' (Bourne in Shaw 1801); Hamstall Ridware (Shaw 1801); Tixall (Clifford 1817); Tamworth (Power 1835!); Catholme and Stapenhill (Brown 1863). An undated manuscript note by Garner tells us that 'Mr Bikie first discovered it seventy years back in a meadow where three rivers meet—Tame, Mease and Trent.'

Aquilegia vulgaris L.   Columbine
Woods and thickets on the limestone and Keuper marl, where it is doubtfully native, and as a garden escape: rare: (C), H, (K–L), (N–P), (R), (Y), (a), (z), 1. 'On Needwood Forest, between the Foxholes and Yoxall, 1791' (Gisborne MS). In 1955 a compact colony of about twenty plants with maroon-coloured flowers was found on a rocky part of Ecton Hill in the Manifold Valley.

Thalictrum flavum L.   Common Meadow-rue
Marshy thickets and wet meadows by rivers and brooks: local in the south, particularly in the Trent basin: (S), W–Z, c–d, (f), g–h, j–k, p, (r), t, 16. 'In a meadow on my farm (Pendeford) in considerable quantity' (Pitt 1794). (Map p 199).

T. minus L.   Lesser Meadow-rue
One unconfirmed record for the Weaver Hills, J. Pickard (1925). The plants were said to have been few and small.

BERBERIDACEAE

Berberis vulgaris L.   Barberry
Occasionally seen in hedgerows, particularly in central Staffordshire, but less common than formerly: (F), (H), (N), R, T–U, W–Z, (b), (e), h, (m–n), (r), (t), (w), 8. Pitt, 1794. Bagnall (1901) mentions Colton and Trysull Dingle as two places where it was abundant in his day.

Mahonia aquifolium (Pursh) Nutt.   Oregon Grape
Frequently naturalised in shrubberies and game coverts, but there are no published records: K–L, R–S, U, X–Y, a–c, e, i–j, n–p, v–w, 28.

NYMPHAEACEAE

Nymphaea alba L.   White Water-lily
Perhaps native in the rivers in past days, before they were polluted, but most of the records are for ornamental waters in parks and

gardens: F, (K), (N), (R), S, W, (X–Z), a, g–i, n, (p), (t), 9. Tamworth, where it was called water-can (Withering 1787).

Nuphar lutea (L.) Sm.   Yellow Water-lily
Rivers, canals, lakes and ponds: frequent: E, (F), (J), K–L, N, S–T, W–b, d–k, n–r, (t), u, w, y, 60. Pitt, 1794.

CERATOPHYLLACEAE

Ceratophyllum demersum L.   Rigid Hornwort
Recorded in the past for 'ditches and lakes' (Dickenson 1798) and said to be frequent in the Trent at Burton (Burton Flora), but the few recent records are for canals (two of them now drained) at Dudley, Forton, Great Haywood and Lichfield: (K), W, Y, (b–c), (e), (g), n, (r), x, 4.

PAPAVERACEAE

Papaver rhoeas L.   Common Poppy
Common on arable land and by roadsides in the south and abundant in the rich sandy soil of the south west: in the north it occurs much more rarely and usually as a stray on waste ground: F–H, K–N, R–S, (T), U–W, Y, a–u, w–x, z, 153. Pitt, 1794. (Map p 199).

P. dubium L.   Long-headed Poppy
The commonest poppy in the centre of Staffordshire: further north it is rarer, though more often seen than *P. rhoeas*: in the south west it is less plentiful than *P. rhoeas*: B, D–P, R–u, w, y, 244. Lichfield (Jackson 1837). (Map p 200).

P. lecoqii Lamotte   Yellow-juiced Poppy
Mill Dale, Alstonfield (Purchas 1883!). The specimen in Bagnall's herbarium has not been recently checked.

P. argemone L.   Prickly Poppy
Sandy cornfields, gravel pits, roadside verges and railway embankments: infrequent: K, (L), (N), R, V, (Y), Z, (b), (f), g, j, m–n, r, w, 17. Dickenson, 1798.

P. somniferum L.   Opium Poppy
Seen occasionally on waste ground as a garden outcast: G, (V), (b), (e), j, (n), s, 3. Lichfield (Jackson 1837). 'Below Tutbury Castle, with the character of the wild plant' (Garner 1844 and Fraser 1884!).

Chelidonium majus L. Greater Celandine
Hedgebanks near houses: B, E–n, r–z, 213. Pitt, 1794. Var *laciniatum* (Mill.) Koch at Alton (Berrisford 1906!). (Map p 200).

FUMARIACEAE

Corydalis claviculata (L.) DC. Climbing Corydalis
In dry gritstone woods, in woods and hedgerows on the Bunter sandstone and by streams in woods bordering peat mosses: B–C, F–H, K–T, (U), W–Z, c–d, f–g, j, n, (t), w–x, 62. First recorded by Ray (1670) for the banks of the Trent near Wolseley, where it was seen on 13 May 1662 in great plenty (Derham 1760). (Map p 200).

C. lutea (L.) DC. Yellow Corydalis
Often seen on garden walls and sometimes persisting for many years on ruins, as at Throwley Old Hall (Allen 1911! and Hall 1940). Garner, 1844.

Fumaria capreolata L. White Ramping-fumitory
Hedgerows at Ellenhall (1944!), conf M. G. Daker; Chatcull (1954!). Earlier records uncertain.

F. muralis Sond. ex Koch subsp *boraei* (Jord.) Pugsl.
Common Ramping-fumitory
Cultivated land: infrequent: L, R–S, Y–Z, (f), n, w, 10. 'Field near Brereton Cross' (Reader 1922!).

F. officinalis L. Common Fumitory
Cultivated fields in the south: J, K, (N), R–S, W–Z, b–g, j–u, w–y, 98. Pitt, 1794. (Map p 200).

CRUCIFERAE

Brassica napus L. Rape
A relic of cultivation recorded for arable fields, river banks and waste ground. Near the Dove (Pitt 1794).

B. rapa L. Wild Turnip
'Much cultivated in light soils' (Dickenson 1798). Garner (1844) said it was less common than *B. napus*. What may be the wild plant has been recorded for the river Trent at Stoke (Garner 1844) and Armitage (Reader 1919!).

B. juncea (L.) Czern.
Casual. Recorded by Druce for Brocton camp, Lichfield and

Burton (BEC 1919, 1923, 1926 Reps); Stoke (1932!); Newcastle, as a bird-seed alien (1970).

B. nigra (L.) Koch   Black Mustard
Perhaps native on the banks of the Severn at Arley (Fraser 1865! and Edees 1954!), but elsewhere recorded recently only for roadsides and rubbish tips: L, (Y), Z, (b), (e), f, (g), m, s, (t), v, 8. Pitt, 1794.

Sinapis arvensis L.   Charlock
Arable fields: B–U, W–z, 520. Dickenson, 1798.

S. alba L.   White Mustard
Relic of cultivation: (F), (L), (N), (R), (T), (Y), b, (f), n, (r), (t), 2. 'Cornfields, frequent' (Dickenson 1798).

Diplotaxis muralis (L.) DC.   Annual Wall-rocket
Walls, sandy roadsides, waste ground and railway banks: rare. Walls of Lichfield Close (Dickenson 1798); garden wall at Stapenhill (Brown 1863); railway bank, Brereton colliery (Reader 1922!); Hoar Cross (BEC 1923 Rep); abundant at Weston on Trent (1932); North Street, Stoke (1932); Newcastle (1933!); railway embankment, West Bromwich (1950); Cock Lane, Bednall (1957!).

D. tenuifolia (L.) DC.   Perennial Wall-rocket
Lichfield Close (Ray 1724 as *Eruca sylvestris*); Kinver Edge, very rare (Bagnall 1869!).

Raphanus raphanistrum L.   Wild Radish
A widespread weed of arable land, but most plentiful in the south, where it sometimes abounds in sandy fields and disused gravel pits: B–P, R–y, 387. Pitt, 1794. The colour forms have not been studied, but a pure colony with deep golden-yellow unveined petals was found in a sandy fallow field near Long Birch Farm, Brewood, in 1954.

Rapistrum rugosum (L.) All.
Casual. Burton (Druce in BEC 1928 Rep and Harlond 1941!); West Bromwich, waste ground in the Hill Top area (1949!); Waterhouses, old railway track (1955!).

Conringia orientalis (L.) Dumort.   Haresear Mustard
Casual. Oakamoor (Berrisford 1902!); Croxden (1917); Alton (1919); Lichfield (BEC 1923 Rep); Burton (BEC 1928 Rep).

Lepidium sativum L.   Garden Cress
Roadsides, railway banks, river banks and rubbish tips: field
records not kept. 'Naturalised in a remote lane at Branston'
(Brown 1863).

L. campestre (L.) R.Br.   Field Pepperwort
Cornfields, railway embankments and roadsides: H, (N), R–S, (X),
Z, (b), (g), (n), (r), (v), 4. Jackson, 1837.

L. heterophyllum Benth.   Smith's Pepperwort
Roadsides, railway embankments and waste ground, particularly
near Newcastle and south of Wolverhampton: K–L, (N), R, (S), V,
(b), r, (w), 5. Garner, 1844.

L. ruderale L.   Narrow-leaved Pepperwort
Casual: frequent on rubbish tips in the south: (K), L, (N), (S), (U),
(Y), (b), e–h, k–p, s–t, 14. 'On some waste ground near the Tean
Brook at Fole' (Goodall 1882).

L. latifolium L.   Dittander
Tixall salt marsh in 1840 (Reader 1922); on waste sandy ground in
an old brick pit at Wolseley Bridge (1964!).

Coronopus squamatus (Forsk.) Aschers.   Swine-cress
Gateways and tracks through fields, especially where the ground
has been manured: more frequent in the east of the county than
elsewhere, but uncommon: (J), S, (U), (Y), (b), f–g, j, p, (x), 8.
Garner, 1844.

C. didymus (L.) Sm.   Lesser Swine-cress
Field entrances, a sand pit, a railway bank, a garden, a cornfield and
a road verge: uncommon: B, K, L, (N), (a–b), (f), g, i, w, y, 9.
Churchill Babington in Watson, 1837.

Cardaria draba (L.) Desv.   Hoary Cress
An increasing species of waste ground in towns, now locally
abundant in the Potteries and Black Country: F, L, (N), T, (Y),
b, m, p, s, x–y, 18. Berrisford, 1902.

Thlaspi arvense L.   Field Pennycress
Cultivated fields and waste ground: frequent in the south: C, L–M,
(N), S, (T), W–Z, (b), d–g, j, (m), n–p, r–s, v–y, 37. 'Stone in
Staffordshire' (Ray 1670).

Teesdalia nudicaulis (L.) R.Br.   Shepherd's-cress
Sandy heaths, in open places where the grass is short: today known
only for the golf course on Highgate Common: (G), (K), (M), (S),
(Y), (d), (f–g), (n), (t), w, 1. 'On the left hand bank, one mile from
Lichfield, road to Walsall, 30 May 1794' (Gisborne MS).

Capsella bursa-pastoris (L.) Medic.   Shepherd's-purse
Arable land and waste ground: A–z, 770. Pitt, 1794. Druce (BEC
1922–30 Reps) records twelve micro-species for Staffordshire.

Hornungia petraea (L.) Reichb.   Hutchinsia
Limestone rocks: rare or local. 'In Dovedale on the rocks, May
1793' (T. Gisborne MS). 'Weaver Hills' (J. Gisborne 1797). It has
often been recorded for Dovedale. Fraser (1864) found it plentiful
on limestone rocks at the entrance to Dovedale and Smith (1871)
said it grew on rocky ledges chiefly on the Staffordshire side of
Dovedale and in Beresford Dale. It is still fairly common on
Bunster Hill, at Wetton Mill, in Hall Dale and in Dovedale at the
entrance to Hall Dale.

Cochlearia danica L.   Danish Scurvy-grass
In the gravel of railway tracks: probably more frequent than our
five records suggest, but as impermanent as the disappearing
railways. Hamstead, West Bromwich, V. Jacobs (1948!); railway
north of Aston Cliff, Maer (1957!); disused railway between
Trentham and Trentham Park (1959!); Cliffe Park near Rushton
(1964!); Endon (1965!).

Alyssum alyssoides (L.) L.   Small Alison
Clover field at Langley, Lower Penn (Fraser 1864!); field near
Loo Mill (Fraser 1876!); Oakamoor (Berrisford 1902!); Weaver
Hills (Berrisford 1903!).

Draba incana L.   Hoary Whitlow-grass
'On limestone rocks by Thor's Cave, pod curiously twisted, rare'
(Garner 1844); Dovedale (Bagnall 1901); 'On Caldon Low, road
above Star inn' (Berrisford MS, undated). These old records have
not been confirmed, but in 1950 I saw a specimen collected by a
school boy at Hollinsclough which was said to have been found in
the garden of Croftbottom Farm.

D. muralis L.   Wall Whitlow-grass
Local on limestone rocks in the valleys of the Dove and Manifold

and on the Weaver Hills and sometimes found on limestone rubble
used for road making: C, H–J, N–P, 13. Gisborne, 1793.

Erophila verna (L.) Chevall.   Common Whitlow-grass
Locally abundant on the limestone hills, where there are outcrops
of rock, and in bare places on heaths in the rich sandy district of
Enville and Kinver: uncommon elsewhere, but recorded for walls,
gravel walks and old mine heaps: (F–G), H–J, (M), N–P, W, Y–Z,
b, (e–f), g, (n), r, w, 38. Pitt, 1794. (Map p 201).

Armoracia rusticana Gaertn., Mey. & Scherb.   Horse-radish
Naturalised on river banks and in many places on waste ground
near houses, but often recorded merely as an outcast from gardens:
B, E–H, K–y, 360. Brown, 1863.

Cardamine pratensis L.   Cuckoo-flower
Lowland water meadows and wet upland hayfields: A–z, 577.
Pitt, 1794. A colony of fifty plants with pure white petals near
Haywood Park Farm on Cannock Chase (1948!).

C. amara L.   Large Bitter-cress
Swampy woods, reed beds and marshes by rivers and canals:
common in Dovedale and in many places throughout the county:
B–D, F–U, W–n, q–r, t–w, y, 173. 'In the bog near the bath at
Willowbridge' (Waring 1770). (Map p 201).

C. impatiens L.   Narrow-leaved Bitter-cress
Limestone woods, rare: a casual elsewhere: H–J, (N), (V), (n), (s),
(v), w, (x), 4. 'Barrow Hill at the quarries of the Rowley Rag and at
Sedgley' (Wainwright in Shaw 1801). There are several records for
Dovedale and the Manifold Valley, the most recent being for
Musden Wood (1951!) and Cheshire Wood (Shimwell 1968).
Abundant near Rocester (1913). In the south west recorded for
cinder heaps at Whittington (Fraser 1877!); Arley (Fraser 1884!);
canal side between the Vine and the Stewponey (1953!).

C. flexuosa With.   Wavy Bitter-cress
Marshy meadows, ditches, walls and boggy places in woods: A–w,
y–z, 534. Dickenson, 1798.

C. hirsuta L.   Hairy Bitter-cress
Walls, dry banks, railways, cinder paths and about rocks: abundant on the Carboniferous limestone and on the sandy heaths in
the south west: elsewhere less common than *C. flexuosa*, but easily

overlooked, as it flowers early and soon withers: B, G–P, R–U, W, Y, a–b, d–e, (f), g, j, n, (r), t, v–w, y, 73. 'Gravelly soil, on the driest banks . . . Staffordshire, common' (Withering 1796).

C. bulbifera (L.) Crantz   Coralroot
Rare and of doubtful status. Pendeford, 'hedge sides on this farm' (Pitt 1796); Blithfield, grove by the churchyard (Garner 1844); Yoxall (1945!); dingle between Trentham Park and New Inn Lane (1958!). These are the only recorded stations and in none of them is the plant far from houses. But it looks native in Needwood Forest on the Keuper marl. I have seen it in three places near Yoxall Lodge, (1) by the stream to the north west of the house, (2) abundantly in a copse between Newchurch and Scotch Hill and (3) in the wood, formerly known as Coalpit Slade, on the east side of the road near Darleyoak Farm. Garner recorded it for Needwood Forest in 1844 but did not quote his authority. There is no proof that Gisborne knew it.

Barbarea vulgaris R.Br.   Winter-cress
River banks, canal sides, damp roadsides and occasionally on ploughed land: common in many places, particularly along the course of the Dove and Trent: B, E–M, (N), P, R–b, d–j, m–z, 130. Near the Dove (Pitt 1794).

B. verna (Mill.) Aschers.   American Winter-cress
Unconfirmed records for Stoke (Garner 1844), Burton, Cheadle and Oakamoor, but I have seen no satisfactory specimens.

Arabis hirsuta (L.) Scop.   Hairy Rock-cress
Common on limestone rocks in Dovedale and the Manifold Valley and frequently recorded for the Weaver Hills: H–J, N, (b), 17. 'Stony places, amongst the ruins of Tutbury Castle' (Dickenson 1798).

Turritis glabra L.   Tower Mustard
Recorded in the past for roadsides and hedgebanks in the sandstone districts of south Staffordshire, but not seen for a hundred years: (c), (n), (r), (w). Lichfield (Whately in Withering 1787) and 'in hedges near Lichfield leading to Tamworth' (Clifford 1817).

Rorippa nasturtium-aquaticum (L.) Hayek sensu lato   Water-cress
Streams and ditches, in clear flowing water: B–C, E–y, 362. Pitt, 1794. The distribution of the segregates has not been worked out,

but most of the plants examined in the field appeared to belong either to *R. microphylla* (Boenn.) Hyland. or *R. microphylla* x *nasturtium-aquaticum*. Of thirty-one Staffordshire specimens in my herbarium five have been named *R. nasturtium-aquaticum* sensu stricto and the rest go about equally to *R. microphylla* and the hybrid.

R. sylvestris (L.) Bess.   Creeping Yellow-cress
Riversides and gardens: rare. Blymhill, lane by side of Dawford Brook (Dickenson 1798); Tamworth (Garner 1844); Trent side at Wetmore (Brown 1863); Arley (Bagnall 1901); Patshull (BEC 1917 Rep); Stoke, garden at Hartshill (1943); damp ground by a pool in Himley Wood (1946!); Newcastle, garden at Clayton (1955!).

R. islandica (Oeder) Borbás   Marsh Yellow-cress
Ditches, river banks and pond margins, in wet mud, sometimes profusely covering the bottom of a pond when the water has evaporated: widespread: B, F–N, S–x, 172. 'Near Stafford, on the road to Castle Hill' (Stokes in Withering 1787). (Map p 201).

R. amphibia (L.) Bess.   Great Yellow-cress
Frequent in the south in ditches, streams, canals and rivers, but in the north almost confined to the canals: G, L–M, (N), P, S–T, (W), Y–Z, b, d–r, t–w, y, 59. 'Side of the river and wet ditches at Tamworth' (Withering 1787). (Map p 201).

Hesperis matronalis L.   Dame's-violet
Garden escape: C, (N), R, (f), (r), (x), 2. Carter, 1839.

Erysimum cheiranthoides L.   Treacle Mustard
Railway sidings and embankments, waste heaps, allotments and riversides: uncommon: G–H, (L), (Y), a–b, e–f, (g), j, (n), r–t, v, y, 19. Fraser, 1864.

Cheiranthus cheiri L.   Wallflower
'On old walls' (Pitt 1794); moat walls at Perry Hall (Pitt 1796); Dudley Castle, Burton Abbey, ruins of Rugeley old church (Garner 1844); Tutbury Castle (Nowers 1901!).

Alliaria petiolata (Bieb.) Cavara & Grande   Garlic Mustard
Hedges throughout the county, except for the hill country, Cannock Chase and some of the industrial areas: E–z, 548. Near the Dove (Pitt 1794). Map p 202).

(*above*) The Roches, the roof of Staffordshire; (*below*) view from Ramshaw Rocks

Thor's Cave, home of the Nottingham Catchfly (*Silene nutans*) and
Jacob's-ladder (*Polemonium caeruleum*)

Sisymbrium officinale (L.) Scop.   Hedge Mustard
Uncommon in the north, but abundant everywhere else in cultivated
fields, by roadsides and on waste ground: E–z, 435. Pitt, 1794. Var
leiocarpum DC. at Burton (BEC 1930 Rep) and Penn (BEC 1941
Rep). (Map p 202).

S. orientale L.   Eastern Rocket
Open waste spaces and rubbish tips in the towns: frequent and
increasing: F, H, K–M, (N), Y, a–b, e–f, k–u, w–y, 43. 'By road
from Croxall to Elford' (Reader 1922!).

S. altissimum L.   Tall Rocket
Waste ground: frequent in the south, rare in the north: L, (N),
(Y), (b), e, k–m, s–t, x–y, 14. Brocton (Druce in BEC 1919 Rep).
Reader (1920!) found it at Milford and Brocton railway station.

Arabidopsis thaliana (L.) Heynh.   Thale-cress
Walls and dry bare places throughout the county, but most common
on the Carboniferous limestone and in sandy fields in the south
west: B–C, (F), G–L, N–Q, S–U, W–Z, b–k, n–r, t–w, y, 127.
Pitt, 1796. (Map p 202).

Camelina sativa (L.) Crantz   Gold-of-pleasure
Cornfields in former days. Cornfield near the parsonage at Blymhill
and several times at Cheddleton (Garner 1844); '1854, abundant in
a cornfield on Penkhull Hills' (Garner MS); Orton near Wombourn
(Fraser 1864); Wetley Rocks (1869); brickyard, Codsall (Fraser
1873!); Whittington, 1877 (Mathews 1884); Perton (Bagnall 1901);
Oakamoor (Berrisford & Walker 1906); fields near Newcastle and
at Dilhorne (Ridge in Flora).

Descurainia sophia (L.) Webb ex Prantl   Flixweed
Casual. Old records for Tutbury Castle (Dickenson 1798); Coton
Field (Garner 1844); Walton Lane (Brown 1863); Burton (Nowers
1894!); Oakamoor (Berrisford & Walker 1906); Stoke (1917).
Recent records for Aldridge, waste ground by the canal near
Wharf Bridge (1957!); Clayhanger, roadside between the flashes
(1957!); roadside near Rugeley railway station (1967).

RESEDACEAE
Reseda luteola L.   Weld
Coal pit mounds, railway sidings and waste ground: long estab-
lished in the industrial areas and one of the most abundant weeds

of the Black Country: (F), K–M, (N), S, (W), X–Y, b, d–g, j–u, w–y, 125. 'Coal pit banks in Staffordshire . . . and about the ruins of Dudley Castle' (Withering 1787). (Map p 202).

R. lutea L.   Wild Mignonette
Waste ground in the industrial areas: (B), K–L, (N), (S), Y, b, f–g, m, p, (s), t, 16. 'Rough ground near Russell's Hall, Dudley' (Pitt 1817).

VIOLACEAE

Viola odorata L.   Sweet Violet
Hedgebanks and woods in base-rich soils: less common than formerly, but plentiful in Needwood Forest: J–L, (M), N–P, R–S, U–k, n–s, v–w, y, 85. 'Warm hedges and ditch banks, and in moist warm lanes, particularly in clay or marl' (Withering 1787). The white-flowered var *dumetorum* (Jord.) Rouy & Fouc. is commoner than the typical plant with violet flowers. Var *imberbis* (Leight.) Henslow (det S. M. Walters) has been recorded for Mitton, Penkridge (1949!). (Map p 203).

V. hirta L.   Hairy Violet
Confined to the Carboniferous limestone, where it is frequent in thickets and on grass slopes about outcrops of rock: H–J, 7. Dovedale (Bree in Purton 1817). *V. hirta* x *odorata* (det P. M. Hall) has been recorded for the Manifold Valley (1938!).

V. riviniana Reichb.   Common Dog-violet
Woods, hedgebanks, heaths and grassy hillsides: A–n, q–z, 519. Dickenson, 1798. Subsp *minor* (Gregory) Valentine occurs on sunny hilltops.

V. reichenbachiana Jord. ex Bor.   Early Dog-violet
A local species of woods and shady hedgebanks in rich soils: frequent in the Manifold Valley and Needwood Forest: H–K, N–P, R, U, W–X, Z–a, (f–g), r, (s), (v), w, 26. Bagnall, 1901, and Nowers, 1901. (Map p 203).

V. canina L.   Heath Dog-violet
'Turfy ground near Stile Cop' (Reader 1923); 'High moorlands near Sherbrook, Cannock Chase' (Reader 1924!); Enville, in two places near the golf course on Highgate Common (1960!); Eccleshall, on an ant hill in a rough pasture at Offleybrook (1960!). These are the latest and most reliable records, but there are earlier

ones, under the name *V. flavicornis*, for Cannock Chase, Enville and Needwood Forest, which may be correct. On 10 June 1837 Babington wrote in his diary: 'I reached Yoxall Lodge. I noticed near the Lodge *Genista anglica, Viola flavicornis, Orchis morio*, and *Myosotis versicolor*, all of them in plenty' (A.M.B. 1897). Three of these are rare plants in Staffordshire today.

V. palustris L. Marsh Violet
Acid swamps near running water, its leaves often marking out the course of a hidden rill: locally plentiful, especially on the moors: B–K, M–P, R–U, W–a, c, e–f, m–n, r, (u), w, 97. 'On Brakenhurst Bog, 1791' (Gisborne MS); 'In most bogs where *Drosera* and *Anagallis tenella* grow' (Forster in Clifford 1817). (Map p 203).

V. lutea L. Mountain Pansy
A plant of the hills, seldom found much below 1,000ft: local: B–D, H–J, N–P, 26. Most plentiful on the upper slopes of limestone hills where the soil is stable, but it also occurs on the flanks of the Roches. 'In some mountainous pastures on the road from Leek to Buxton' (Forster 1796). Gisborne (1797) found it on the summits of the Weaver Hills and wrote, 'Many of the mossy knolls and hillocks are frequently studded with the golden petals of this elegant plant.' Yellow-flowered plants predominate in Staffordshire. (Map p 203).

V. tricolor L. Wild Pansy
Cornfields and sandy roadsides: infrequent: H, N–P, U, W, (Y–Z), (f), j, (m–n), r, u, (v), 10. 'Roadsides about Wolseley Bridge, in a loose sandy gravel. . . . Its elegant blue flowers are a grateful ornament to the barren soil in which they grow' (Stokes in Withering 1787). Withering (1801) refers to these plants again as being very fine in the lanes about Bishton. P. M. Hall named a yellow pansy found in the lane between Warslow and Ecton in 1937 *V. lepida* Jord.

V. arvensis Murr. Field Pansy
Arable fields: common in the south: (F), K–L, N, R–U, W–u, w–y, 207. The commonest segregate is the plant we used to call *V. segetalis* Jord. f. *obtusifolia* (Jord.) Drabble. Specimens from Needwood Forest, collected by Babington in 1832, were determined by Drabble (BEC 1936 Rep 321) as *V. arvatica* Jord., *V. deseglisei* Jord. and *V. derelicta* Jord. There are fifty-two Staffordshire specimens of the aggregate species in my herbarium. (Map p 204).

POLYGALACEAE

Polygala vulgaris L.   Common Milkwort
Base-rich pastures and limestone hillsides: C, E, G–K, N–P, R–S,
W, b, (f), s, 27. 'Mountainous heathy pastures near Cheadle;
limestone hills, Sedgley' (Pitt 1796).

P. serpyllifolia Hose   Heath Milkwort
Heaths and moors: common on Cannock Chase and in the north
of the county: B–C, E–H, K–N, R–T, Y–Z, (a–b), e, (f), (m), r, (v),
z, 39. Whitmore (Fraser 1873!).

GUTTIFERAE

Hypericum androsaemum L.   Tutsan
Known today only for Needwood Forest and Seckley Wood.
Pendeford, 'in hedges and waste baulks on this farm' (Pitt 1796);
Needwood Forest (Riley in Shaw 1801), confirmed for Forest
Banks, 'a single plant almost hidden in a tangle of bramble and
rose' (1949); woods near Burslem (Turner & Dillwyn 1805);
Hopwas (Jackson 1837); Leycett and Upper Arley (Garner 1844);
Seckley Wood (Fraser 1874!), seen again in 1954; Enville (Bagnall
1901).

H. calycinum L.   Rose-of-Sharon
Naturalised in Seckley Wood (1954!).

H. perforatum L.   Perforate St. John's-wort
Woods, gravel pits, railway embankments and dry roadsides
throughout the county: G–z, 189. Pitt, 1796. (Map p 204).

H. maculatum Crantz   Imperforate St. John's-wort
Hedgerows, riversides and wood borders throughout the county:
B, E, G–U, W–a, (b), c–w, y–z, 167. Stretton (Grove in Dickenson
1798). All thirteen specimens in my herbarium, which come from
widely separated parts of the county, appear to be subsp *obtusius-
culum* (Tourlet) Hayek.

H. tetrapterum Fr.   Square-stalked St. John's-wort
Common in marshy places: B–D, F–w, 238. Dickenson, 1798.

H. humifusum L.   Trailing St. John's-wort
Lawns, sandy heaths, sand and gravel pits, walls, disused railway
tracks, woodland rides and dry roadside banks: frequent, except
on the limestone: B–C, (F–G), H, K–L, (N–P), R–U, X–a, (b), (d),
e–g, j, n, r, (w), z, 29. Near Yoxall Lodge (Gisborne 1791!).

H. pulchrum L. Slender St. John's-wort
Frequent in heathy woods and bushy places where gorse grows,
particularly on the Bunter sandstone near Eccleshall and in the
Churnet Valley: B–D, (F), G–H, K, (M), N–P, R, U, W–a, (b),
c–e, (f), k, r, v–w, z, 42. 'Blymhill, in fields called the Small Heaths'
(Dickenson 1798). (Map p 204).

H. hirsutum L. Hairy St. John's-wort
Woods on basic soils: common in the Manifold Valley and
Dovedale, in Needwood Forest and in the woods near Arley:
C, H–J, N–P, U–V, (X–Y), a, (b), (t), v, 35. 'Near Tutbury in
hedges' (Dickenson 1798). (Map p 204).

H. montanum L. Pale St. John's-wort
Oakedge near Wolseley Bridge (Clifford 1817), an unlikely place
for a lime-loving plant; Burton (Garner 1844); Dovedale, in small
quantity (Purchas 1885); by the roadside between Wetton Mill and
Wetton (1945!).

H. elodes L. Marsh St. John's-wort
Very rare in acid bogs and now possibly extinct. Needwood Forest
(Gisborne 1791!); Calf Heath (Pitt 1796); near White Sitch Pool
(Dickenson 1798); near Stone (Clifford 1817); near Lichfield
(Jackson 1837); 'In all our bogs and mosses' (Garner 1844);
Chartley Moss (Brown 1863 and Bagnall 1894!). Bagnall (1901)
queries Garner's statement and it is hard to believe. Fortunately
the two surviving specimens, the one representing the earliest
record and the other the latest, are excellent examples of the species.

CISTACEAE

Helianthemum chamaecistus Mill. Common Rock-rose
Limestone pastures: local: H–J, N–P, 15. 'Near Thor's House
Cavern' (Pitt 1796).

ELATINACEAE

Elatine hydropiper L. Eight-stamened Waterwort
Unconfirmed records for two pools on Cannock Chase. Pottal
reservoir (Bagnall 1895!); on mud of dried up pond, Slitting Mill,
near Rugeley (Reader 1923!). Bagnall (1895) wrote: 'The petals
were a beautiful pale rose colour; stamens eight; capsules sessile,
containing a few ovules curled like a horse-shoe, with unequal
sides; the plant formed coral-like tufts.' In 1901 he said it was
abundant.

CARYOPHYLLACEAE

Silene vulgaris (Moench) Garcke   Bladder Campion
Cornfields, 'marly banks . . . and in railway cuttings' (Brown 1863), pit mounds, limestone quarries, railway sidings and waste places in town areas: widespread, but most common in the Black Country: C, F–T, W–g, j–n, r–u, w–z, 169. 'Often growing amongst barley on light land' (Pitt 1794, p. 95).

S. gallica L.   Small-flowered Catchfly
Sandy fields and waste ground: rare. 'In cornfields at Upper Arley' (Garner 1844); field near Enville Common (Fraser 1865!); railway cutting, Streetly (Bagnall 1901); Churnet Valley railway (Allen 1910!); garden weed at Bar Hill, Madeley (1925); Burton (BEC 1930 Rep); edge of cornfield half-way between Seisdon and Hillend (1954!).

S. nutans L.   Nottingham Catchfly
Outcrops of the Carboniferous limestone: local. Dovedale, 1792 (Gisborne MS). Recent records for Thor's Cave, Bunster Hill, Mill Dale, Gipsy Bank and the northern tip of Gratton Hill.

S. noctiflora L.   Night-flowering Catchfly
A rare casual of arable fields and waste ground. Lichfield (Jackson 1837): Abbots Haye (Masefield 1884!); Oakamoor (Berrisford 1902!); Alton (1918); Hawkesyard (Reader 1923); Burton (Curtis 1930); Caldonlow quarries (1957).

S. dioica (L.) Clairv.   Red Campion
Throughout the county in woods and hedgerows: A–r, t–w, y–z, 638. Woodmill meadows (Gisborne 1790!).

S. alba (Mill.) E. H. L. Krause   White Campion
Fallow fields and open sunny places in light soils: frequent in the north, common in the south: C, E, G–N, Q–S, U–Y, 293. Dickenson, 1798.

Lychnis flos-cuculi L.   Ragged Robin
Marshy places in lowland meadows and by upland streams, alike in rich and poor soils: common: A–U, W–r, t–w, y–z, 358. Woodmill meadows (Gisborne 1791!).

Agrostemma githago L.   Corncockle
Cornfields: once frequent, if not common, but now rare: B, (F–H), (L–N), (R), S, (Y), (b), (f–g), (t), 2. Pitt, 1794. 'Common, looking

handsome among corn, but not desired by the farmer' (Garner 1844); Rushton, 'about a dozen plants along the edge of a wheat field' (1944); barley field near Nelson Hall (1947). (Map p 205).

Vaccaria pyramidata Medic. Cowherb
Casual. Burton (Nowers 1889! and later records); Oakamoor (Berrisford 1902!); Patshull (BEC 1923 Rep); Mayfield, one plant in a farmyard (1955).

Saponaria officinalis L. Soapwort
Garden escape: L, (T), (Y–Z), (b), (d), (f–g), (p), s, (v), x–y, 6. 'In a ditch bank at ... Hamstall Ridware' (Riley in Shaw 1801).

Cerastium arvense L. Field Mouse-ear
Sandy heaths and tracks, chiefly in the south west: rare. Kinver Edge (Fraser 1873!); dry hillsides, Hawkesyard Park (Reader 1920!); Highgate Common and Whittington Common (1960!); Patshull, a few plants by one of the tracks in the north east corner of the park (1961!). Garner (1844) said it could be found occasionally on limestone, but this has not been confirmed.

C. tomentosum L. Snow-in-Summer
Well established in disused quarries at Caldonlow (1957).

C. holosteoides Fr. Common Mouse-ear
Cultivated fields and waste ground: A–z, 780. Yoxall Lodge (Gisborne 1787!).

C. glomeratum Thuill. Sticky Mouse-ear
Arable fields and waste ground: widespread, but less plentiful than *C. holosteoides:* B–r, t–w, y–z, 269. 'Under Swilcar Oak' (Gisborne 1792!).

C. diffusum Pers. Sea Mouse-ear
Railway tracks:B, M, S–T, X, t, 8. First recorded in 1941.

C. semidecandrum L. Little Mouse-ear
Shallow dry usually base-rich soils: recorded for the Carboniferous limestone, roadsides and ant hills on Cannock Chase and the sandy heaths of the south west: H–J, N–P, Y, f, (m–n), r, w, 11. Highgate Common (Fraser 1865!).

Myosoton aquaticum (L.) Moench Water-chickweed
River gravel and wet places by streams and lakes: rare in the north, common in the south: H, L, S–b, d–h, k, n–t, w, y, 86. 'At Burston

and in the lane leading from the turnpike road to Gayton' (Forster in Clifford 1817). (Map p 205).

Stellaria nemorum L.   Wood Stitchwort
In wet woods by streams: very local. Walton's Wood, Madeley (J. Daltry 1840!); between Alton Towers and Oakamoor (Bagnall 1887!); wood by stream near North Longdon (Murray 1920!); Ravensclough Wood (1943!). It was often recorded for the Churnet below Alton Towers and is still plentiful there on the south side of Lord's Bridge.

S. media (L.) Vill.   Common Chickweed
Gardens, ploughed fields and waste ground: A–z, 785. 'Abounds most in land rendered fine by repeated ploughings' (Pitt 1794). 'It may be worthy of remark that chickweed is an excellent out-of-door barometer: when the flower expands boldly and fully, the farmer need not be apprehensive of rain for four hours and upwards: if it continue in that open state no rain will disturb the summer day: when it half conceals its miniature flower, the day is generally showery: but when it entirely shuts up, or veils the white flower with its green mantle, let the traveller put on his great coat, and the farmer, with his beasts of the plough, rest from their labour' (Shaw in Pitt 1796, p. 65).

S. neglecta Weihe   Greater Chickweed
Damp shady hedgebanks and wood borders: frequent: B, J–L, N–P, R–U, W–g, j–k, r, w, 45. Branston (Nowers 1889!).

S. holostea L.   Greater Stitchwort
Woods and hedgebanks: A–z, 580. Pitt, 1794.

S, palustris Retz.   Marsh Stitchwort
Unconfirmed records for Tatenhill (Acland in Purton 1821); Compton and Wightwick meadows (Fraser 1864); Kinver Edge (Bagnall 1901); Patshull (BEC 1917 Rep); marshy field near Pipe Ridware (Reader 1920!); Lichfield (BEC 1931 Rep).

S. graminea L.   Lesser Stitchwort
Dry heathy places and wet rushy meadows: B–w, y–z, 530. Dickenson, 1789.

S. alsine Grimm   Bog Stitchwort
Marshy meadows, damp woodland rides and by moorland streams: A–w, y, 545. 'In a bog near the Hollyfalls' (Gisborne 1792!).

Moenchia erecta (L.) Gaertn., Mey. & Scherb.   Upright Chickweed
No recent records. Lichfield race ground (Acland in Purton 1821);
near Woodmill, 1837 (Babington 1897); sandy commons, Barlaston
(Garner 1844); Pond Green, Seckley Wood (Fraser 1866!);
Sherbrook and Abraham valleys, Cannock Chase (Bagnall 1901).

Sagina apetala Ard.   Annual Pearlwort
Walls, gravel paths, railway tracks: frequent: G–H, K–M, R–U,
b–c, e, (f), g–j, (m), p–r, (v), 29. Dickenson, 1798.

S. ciliata Fr.
In Staffordshire as in Derbyshire (see Clapham 1969) this species
shows a preference for the Carboniferous limestone, where it grows
with *Minuartia hybrida* on rock ledges. Elsewhere it can be found
in the same habitats as *S. apetala* but is rarer: J, L, S, (f), r, (w),
(z), 8. Seckley Wood (Fraser 1866!).

S. procumbens L.   Procumbent Pearlwort
Damp walls, garden paths, lawns, pavements and stony places in
meadows by streams and ponds: A–z, 577. Pitt, 1796.

S. nodosa (L.) Fenzl   Knotted Pearlwort
Damp peaty places on moors and commons, on the margins of
lakes and in the stonework of canal locks: infrequent and seldom
more than a few plants in any one place: C, G–H, (J), (L), N, W,
(e), m–n, (r), 10. Dovedale (Gisborne 1792!). It has been known on
Ipstones Edge since 1921.

Minuartia verna (L.) Hiern   Spring Sandwort
'Upon fragments of spar near lead mines, Ecton Hill, Dovedale'
(Garner 1844). Rare in both localities. In 1885 Purchas remarked
on its absence from all the mine hillocks which he had examined
on both sides of Dovedale and it is not recorded for Dovedale in
any of the Derbyshire Floras. However, there is a specimen in
Bagnall's herbarium with the label 'Dovedale, Staffs, June 1890'.
Garner withdrew the Ecton Hill record in 1878, but in 1968
Shimwell found it on two of the spoil heaps and in 1970 a single
fine plant was seen in full flower on a path near the river at the foot
of the hill.

M. hybrida (Vill.) Schischk.   Fine-leaved Sandwort
Rocky limestone slopes: rare. Weaver Hills (Carter 1839); Dovedale
(Garner 1860); Cauldon (1866); Froghall (Ridge 1922); Wetton

Hill and Hall Dale (1947!). Purchas (1885) complained that in 1879 all the plants which used to occur on one of the rocky banks near the path in Dovedale 'had been taken away by some ruthless and inconsiderate plant collector.' Alas!

Moehringia trinervia (L.) Clairv.　　　　Three-nerved Sandwort
Woods and hedgerows: B–k, n–r, u–w, 328. Dickenson, 1798. (Map p 205).

Arenaria serpyllifolia L.　　　　　　Thyme-leaved Sandwort
Walls and dry bare ground on sandstone and limestone: to be found in many parts of the county, but, as Garner (1844) said, 'most common on limestone': B–C, (F), G–L, N–P, R–S, (T), U–W, Y–Z, b, d–n, r–x, 105. Dickenson, 1798. (Map p 205).

A. leptoclados (Reichb.) Guss.　　　　Slender Sandwort
Stubble fields, roadside banks, railway tracks and waste ground, in dry stony soil: uncommon: G, X, Z, b, (f), g, (j), p, r, w, 11. Bagnall, 1901.

A. balearica L.　　　　　　　　Mossy Sandwort
Handsworth, established in turf in the cemetery, W. H. Hardaker (1958); Whitmore, in a wild part of a garden at Shutlanehead (1966!).

Spergula arvensis L.　　　　　　　Corn Spurrey
Cultivated fields: A–C, E–z, 444. 'Common on this farm (Pendeford) on some poor arable land when in tillage' (Pitt 1794). Garner (1844) spoke of it 'choking the crops in wet fields.' Var *arvensis* with papillose seeds and var *sativa* (Boenn.) Mert. & Koch having narrowly winged seeds without papillae are both common and often grow intermixed.

Spergularia rubra (L.) J. & C. Presl　　Common Sandspurrey
Sandy heaths, sandy fallow fields and gravel walks: frequent on Cannock Chase and about Kinver and Enville, but infrequent elsewhere: L, (N), Q–S, (T), U, Y–Z, (a–b), e–g, k, (m), n–r, w, 31. Near Yoxall Lodge (Gisborne 1791!).

S. marina (L.) Griseb.　　　　　　Lesser Sandspurrey
Salt marshes near Stafford. All the records come from grid square SJ/92 and because of the special interest of this species, which may not survive the draining of the fields, they are given in full. 'In a salt marsh near Shirleywich' (Stokes in Withering 1787); Pasture-

fields, between the river and the canal, SJ/992248 (1947! and 1956!). 'Abundant in a salt marsh near Kingston' (Garner 1844); field above Kingston Pool (Fraser 1865! and 1884!); still at the marsh by the site of Kingston Pool (Reader 1923! and 1924). 'Rickerscote, covering the ground' (Garner 1844). Muddy remains of salt marsh, Stafford to Baswich (Reader 1923!). Remains of salt marsh north of Stafford, SJ/923259 (1961!).

ILLECEBRACEAE

Herniaria cinerea DC.
Railway sidings and waste ground, Burton (Druce and Curtis in BEC 1926 Rep): still there in 1970.

Scleranthus annuus L. Annual Knawel
Frequent in sandy arable fields south west of a line from Audley to Tamworth: (F–G), (K), R–S, W–X, (Y), Z, (b), d–f, j–n, r, t, w, 41. 'Cornfields on this farm' (Pitt 1796). (Map p 206).

PORTULACACEAE

Montia fontana L. Blinks
Wet flushes and spring heads on sand and peat: often recorded for the Leek moors and Tannock Chase: B–D, G–H, K, M–P, *X*–S, X, (Y), c, e–f, (m), (r), y, 42. 'Blymhill, in a lane by the side of Shallow-ford brook' (Dickenson 1798). Two subspecies have been recognised in Staffordshire, viz subsp *amporitana* Sennen and subsp *variabilis* Walters, but the majority of the specimens have not been critically examined.

M. perfoliata (Willd.) Howell Spring-beauty
Uncommon as a weed in gardens and orchards: (G), (M–N), P, (S), T, Y, f–g, r, (t), 8. Endon (Edwardes 1878).

M. sibirica (L.) Howell Pink Purslane
Shady hedgebanks, woods and streamsides, in sandy soil: locally common in the north, but rare in the south: B, (F), G, J–L, (M), N–P, R–U, W, Y–Z, (n), (t), 33. 'Naturalised on a bank near Penkhull' (Garner 1844). It was not until the early years of the twentieth century that it began to attract much attention as a wild flower. Then it seems to have increased quickly and by 1911, according to Ridge, had become abundant in the neighbourhood of Maer, Hanchurch, Beech and Knypersley.

CHENOPODIACEAE

Chenopodium bonus-henricus L. Good-King-Henry
Waste ground in villages, especially about farms and in churchyards: frequent: (G), H–J, (L), N–P, S–U, X–c, (d), (f), k–n, (r), s,
x–z, 34. Pitt, 1794.

C. polyspermum L. Many-seeded Goosefoot
Rare or overlooked. Anslow (Brown 1863) and other old records
for Burton, Himley Wood, Enville and Elford. Recent records for
Thorpe Constantine, abundant in a vegetable garden near the
church (1957!); Blithfield (1959!); Barton, rubbish dump along
Bar Lane (1962).

C. album L. Fat-hen
Arable land and waste ground: B–z, 641. Pitt, 1794.

C. opulifolium Schrad. ex Koch & Ziz Grey Goosefoot
Burton (Druce & Curtis in BEC 1926 Rep); waste ground, West
Bromwich (1948), det J. P. M. Brenan.

C. murale L. Nettle-leaved Goosefoot
Records for Aldridge, Burton, Enville, Oakamoor, Tatenhill,
Tixall and Wrinehill, but the specimens have not been critically
examined. Perhaps the first reliable record is that of Druce and
Curtis for Burton and Aldridge (BEC 1926 Rep).

C. hybridum L. Maple-leaved Goosefoot
Miles Green, Audley, G. J. V. Bemrose (1945).

C. rubrum L. Red Goosefoot
Old and unconfirmed records for Burton (Brown 1863), Himley,
Gornal Wood, Stoke and Patshull.

Atriplex patula L. Common Orache
Arable land and waste ground: A–z, 656. Dickenson, 1798.

A. hastata L. Spear-leaved Orache
Cultivated fields, rubbish tips and waste places: less common than
A. patula, but plentiful in the south, especially about Stafford and
on the eastern fringe of the county between Uttoxeter and
Tamworth: F, L, (N), P, S–V, X–b, d–u, w, y, 105. 'A common
weed in kitchen gardens, called John-a-neal by the vulgar, who
sometimes boil and eat it as it greens' (Dickenson 1798).

TILIACEAE

Tilia platyphyllos Scop.   Large-leaved Lime
'Planted about Burton' (T. Gibbs in Burton Flora 1901) and in other places, but native in limestone woods. Purchas (1885) found 'one or two scrubby bushes' of this species or the common lime, he could not say which, at an elevation of over 1,000ft on one of the cliffs of Dovedale. It also occurs in Hinkley Wood near Ilam (Pigott 1969).

T. cordata Mill.   Small-leaved Lime
Here and there as a planted tree, but native in Needwood Forest (Marchington Cliff, Buttermilk Hill, Holly Bush Park, Woodmill, Brakenhurst) and on the Carboniferous limestone (Deepdale south of Grindon and Hinkley Wood). 'Frequent and fine in Needwood Forest, and also in the south of the county' (Garner 1844).

T. cordata x *platyphyllos* (*T.* x *vulgaris* Hayne)   Lime
This is the well known planted tree: B–y, 260. 'A timber tree, common in groves and plantations' (Pitt 1796).

MALVACEAE

Malva moschata L.   Musk Mallow
Dry sunny roadsides, especially in rich soils: frequent: H–P, R–U, W–Z, (b), d–g, j–r, (s), (v), w–z, 67. Dudley Castle (Withering 1787). (Map p 206).

M. sylvestris L.   Common Mallow
Roadsides and waste places: much commoner in the south than the north: E–G, J–K, M–N, (R), S–u, w–y, 205. Pitt, 1794. (Map p 206).

M. neglecta Wallr.   Dwarf Mallow
'Roadsides and villages' (Dickenson 1798) and 'on waste ground and by outbuildings and walls' (Ridge in Flora): (F), K, (L–M), N, S–U, W–b, d–e, (f), g–k, n–p, (r), t, v–w, z, 40. Kiddemore Green near Brewood (Dickenson 1798).

LINACEAE

Linum usitatissimum L.   Flax
Casual: odd plants by the sides of canals and newly made roads: formerly more common as a relic of cultivation: (G), (K), L,

(M–N), (a–b), (d), g, (j), k, p, r, 6. 'Much cultivated' (Dickenson 1798).

L. anglicum Mill.   Perennial Flax
Casual. Whittington Heath near Kinver (Fraser 1877!); Aldridge (Bagnall 1901).

L. catharticum L.   Fairy Flax
Railway tracks, canal sides, limestone quarries and other open habitats in dry shallow usually calcareous soils: C–D, G–P, S, (T), U, W–b, d–e, (f), g, m, r–t, w, z, 102. Pitt, 1794. 'In every hilly pasture' (Garner 1844). (Map p 206).

Radiola linoides Roth   Allseed
'In a sandy soil, frequent' (Dickenson 1798). But the only records are for Barlaston Common and Offley Hay (Garner 1844) and these have never been confirmed.

GERANIACEAE
Geranium pratense L.   Meadow Cranesbill
Riversides and riverside meadows: abundant on the limestone and locally plentiful by the Trent and Severn: C–D, (F–G), H–J, (L), N–P, S–V, Y, a–b, g–h, p, v, z, 63. Dickenson, 1798. (Map p 207).

G. endressii Gay   French Cranesbill
Naturalised near Patshull (Druce in BEC 1922 Rep).

G. phaeum L.   Dusky Cranesbill
Naturalised on hedgebanks near houses in a few places: (N), P, (R), S, (a), g, j, (x), 4. Yoxall Lodge (J. Power 1790s!). Persisted for many years at Darlaston (Ridge 1915! and Edees 1934!), until the site was destroyed; Kiddemore Green (1961!); Ellastone (1965); churchyard at Hamstall Ridware (1969).

G. pyrenaicum Burm. f.   Hedgerow Cranesbill
Hedgebanks here and there: (B), (K), N, (T), U, W, (Y), (b), f, (g), j, n, r, v, (w), 9. Lichfield (Jackson 1837).

G. columbinum L.   Long-stalked Cranesbill
Frequent on limestone rocks in the Manifold Valley and Dovedale and formerly recorded for sandy and gravelly soils in the south: H–J, (N), (r), (t), (w–x), 5. Barr Beacon (Withering 1796); 'Trysull in sandy lanes' (Fraser 1864!).

G. dissectum L.   Cut-leaved Cranesbill
Cultivated fields, roadsides and waste places: B, D, G–N, Q–b, d–s, 128. Near Yoxall Lodge (Gisborne 1791!).

G. molle L.   Dovesfoot Cranesbill
Cultivated fields and open habitats in dry shallow soils, becoming abundant in calcareous districts: E, (F), G–Z, b–w, (y), z, 209. Dickenson, 1798. (Map p 207).

G. pusillum L.   Small-flowered Cranesbill
Arable fields and waste ground in dry sandy soils and about outcrops of rock on the limestone hills: often growing with *G. molle* but less common: (G), H–K, (N), P, R, (S–T), U, W–Z, b, e, (f), g, j, n, q–r, w, 45. Forster in Clifford 1817.

G. lucidum L.   Shining Cranesbill
Shady rocks, old walls and hedgebanks: abundant on the Carboniferous limestone, but elsewhere rather rare and often either introduced or an escape from a garden: C–D, (G), H–J, N–P, S–U, X, (Y), (a–b), c, (e), (m–n), (t), (x), 37. 'Very rare in the southern part of the county, Alton Castle, the ruins of Croxden Abbey, Moorlands' (Dickenson 1798).

G. robertianum L.   Herb-Robert
Woods, hedgebanks, walls and quarries: very common in stony thickets on the limestone: A–s, u–z, 487. 'Road from Lichfield to Stafford, a little beyond the 4th milestone, plentifully' (Saville in Withering 1787).

Erodium maritimum (L.) L'Hérit.   Sea Storksbill
Short turf and bare places on sandy heaths: found recently only on Highgate Common and Cannock Chase, but formerly more frequent, especially in the south west. 'Sandy commons between Enville and Bewdley' (Hunter in Withering 1787). Fraser first found it on Highgate Common at Camp (Hill) Farm (1875!) and it has been seen recently near Highgate Farm (1954!). On Cannock chase it occurred in profusion in Oakedge sande pits (Nowers 1889!) and was rediscovered at Seven Springs in the same locality in 1952.

E. cicutarium (L.) L'Hérit.   Common Storksbill
Sandy fields, particularly between Wolverhampton and Arley, and on limestone rocks in Dovedale: G, J–L, (N), R–S, W, Y, b–c, (d),

e–g, j, n, r, (t), v–w, 48. 'Cornfields in a light soil; Blymhill, in the Pye Hill' (Dickenson 1798). (Map p 207).

OXALIDACEAE

Oxalis acetosella L.   Wood-sorrel
'Woods and hedgebanks' (Pitt 1794): A–g, j–n, r–z, 377.

BALSAMINACEAE

Impatiens noli-tangere L.   Touch-me-not Balsam
'In one or two spots in woods, profusely covering the ground, but originating from gardens; Ramsdell' (Garner 1844); near Maer and at Barlaston (Ridge in Flora). These records have never been confirmed.

I. capensis Meerb.   Orange Balsam
Abundant near the mill at Great Haywood, where it was first recorded for the county by P. P. Thornton in 1941, and in many places along the canals between Great Haywood and Tamworth: Y, Z, b, f–g, n, 12.

I. parviflora DC.   Small Balsam
Arley Wood (Fraser 1865!) and by the Severn at Upper Arley (1954!); Lichfield (BEC 1931 Rep).

I. glandulifera Royle   Indian Balsam
Riverbanks and gardens: G, L–M, S–U, Y–Z, f–g, j, m–t, w–y, 71. Stafford, in a garden (Moore 1889!); Patshull (Lady J. Legge in BEC 1923 Rep); 'Trysull in great abundance and completely naturalised' (Curtis & Druce in BEC 1925 Rep); Tamebridge, abundant (BEC 1926 Rep). These are the earliest records and from them we can deduce that the concentration of this species east and west of the Black Country is at least fifty years old. Its appearance in the north, where it has densely colonised the Lyme Brook at Newcastle, and its spread along the whole course of the Trent is probably more recent, because neither Ridge nor Reader mentions it. (Map p 207).

ACERACEAE

Acer pseudoplatanus L.   Sycamore
Woods and hedges and about hill farms: A–z, 761. Pitt, 1796.

A. campestre L.   Field Maple
Hedgerows and woods, growing best in base-rich soils: E, H–P,

(*above*) Pannierspool Bridge, Three Shires Head; (*below*) Cannock Chase,
where the Hybrid Bilberry (*Vaccinium myrtillus* x *vitis-idaea*) grows

(*above*) The canal at Great Haywood, where the Orange Balsam (*Impatiens capensis*) grows; (*below*) the Essex Bridge over the Trent at Great Haywood

R–z, 397. Pitt, 1796, though Plot (1686, p. 225) refers to a maple with striped leaves in the gardens at Ingestre. 'Very fine in the remains of Needwood Forest' (Garner 1844).

HIPPOCASTANACEAE

Aesculus hippocastanum L.   Horse Chestnut
Planted throughout the county: A–y, 319. Daltry, 1908!

AQUIFOLIACEAE

Ilex aquifolium L.   Holly
Woods and hedgerows: A–z, 739. Plot, 1686. 'Needwood Forest, in great abundance, has been nursed up and encouraged in growth, I suppose, as winter provender for the deer' (Pitt 1794).

CELASTRACEAE

Euonymus europaeus L.   Spindle-tree
Thickets and hedges, preferring base-rich soils: rare: (H), P–Q, (W), (b), d, f, j, v, 8. Hedge at Stretton (Grove in Dickenson 1798). Purchas (1885) found it 'in very small quantity' in Dovedale. Recent records for Brockhurst Coppice (1954!); Cawarden Springs (1956!); Tyrley Wharf (1957); Barrowhill near Rocester (1969); Patshull; woods near Upper Arley.

RHAMNACEAE

Rhamnus catharticus L.   Buckthorn
Thickets on the limestone, uncommon elsewhere: H–J, (N), (T), Z, d, (f), g, (j), (m), (r), t, 7. Pitt, 1794.

Frangula alnus Mill.   Alder Buckthorn
Local in damp sandstone woods, especially on the borders of peat mosses: E–F, (G), K–N, (P), R–T, W, Y–Z, d, f, (g), k–n, r, t, (y), 39. 'In the coppice, called the park, at Willowbridge, Staffordshire, where there is a good deal of it in a very dry gravelly soil' (Waring 1770). There in 1942! (Map p 208).

LEGUMINOSAE

Genista tinctoria L.   Dyer's Greenweed
Rough pastures: frequent in the north, but almost absent from the south: C, F–H, K, M, (N–P), (R), S–T, (U), (W), X, Z–a, (b), (z), 18. Waring, 1770. (Map p 208).

G. anglica L.   Petty Whin
Decreasing: recorded for moist heaths in many parts of Staffordshire in past days, but seen recently only at Stanton (1938!) and
Archford Moor (1955!): (G–H), J, (K–L), (N), P, (R), (Y), (a),
(d), (f), (w), 2. Needwood Forest (Pitt 1794). Babington found it
near Yoxall Lodge in 1837 (A.M.B. 1897).

Ulex europaeus L.   Gorse
Rough pastures and heathy places, usually in lime deficient soils,
but covering the slopes of limestone hills in some parts of the
Manifold Valley: A–z, 530. Barr Beacon (Withering 1787). 'The
burning of a large tract of furze on Weaver Hills and Calton Moor
makes a grand appearance amidst the gloom of night' (Gisborne
1797).

U. gallii Planch.   Western Gorse
Less common than *U. europaeus*, except in the north, centre and
south west: A–C, F–P, R–U, W–a, c, e–g, k–n, r–u, w–x, z, 228.
Barr Beacon (Stokes in Withering 1787). (Map p 208).

Sarothamnus scoparius (L.) Wimm. ex Koch   Broom
Heathy places, gravel pits and railway embankments on the
sandstone: A–B, E–H, K–U, W–g, i–n, r–u, w–y, 281. 'Dry pastures'
(Pitt 1796). 'An improved state of agriculture has almost extirpated
this handsome shrub. About thirty or forty years ago there was
scarcely a farm in the ryeland parts of the county but had a field or
two entirely over-run with tall plants of it, five or six feet high,
which furnished a large supply of oven fuel and materials for
making beesoms, etc. At present a field of broom is a very rare
sight, if at all to be seen in Staffordshire' (Dickenson 1798). 'Fields
where coal has been got over-run with this plant near Wednesbury
and Tipton' (Shaw 1801). (Map p 208).

Ononis repens L.   Common Restharrow
Dry grassy roadsides: frequent: (F), H, L, (N), R–T, W–b, d–g,
j–r, t, v–x, 41. Himley (Pitt 1794). (Map p 209).

O. spinosa L.   Spiny Restharrow
I have seen it only near Syerscote Manor, Thorpe Constantine, on
the Keuper marl, by the side of a path through a ploughed field
(1946!). The other records are not supported by specimens:
'Pastures and hedgebanks' (Dickenson 1798); Stretton and
Rolleston (Brown 1863); near road from Longton to Stone (1911);

Brindley Heath, at junction of ride from Birches Valley by fire lookout (1968).

Medicago falcata L. Sickle Medick
Casual: waste ground in town areas: E, L, (N), (Y), b, s–t, y, 6. Oakamoor (Berrisford 1905!).

M. sativa L. Lucerne
Relic of cultivation: K–L, (N), (b), g, k, (m), n, s, (t), w, y, 11. Pitt, 1794.

M. lupulina L. Black Medick
Fields, roadsides and waste places: B–z, 430. 'Cultivated for permanent pasture' (Pitt 1794).

M. polymorpha L. Toothed Medick
Casual: gardens and waste ground: (H), (N), b, (f), (r), 1. Wightwick (Fraser 1873!); Manifold Valley (Ridge 1914!); Hawkesyard (Reader 1922!); Burton (1944!) and other records.

M. arabica (L.) Huds. Spotted Medick
Casual. Garden weed, Stafford (Moore 1889!); Burton (BEC 1928 Rep); Madeley, in grass at Lower Mill (1958!).

Melilotus altissima Thuill. Tall Melilot
Waste ground and rubbish dumps: rare. Stafford Castle (Fraser 1864!); Codsall (Fraser 1865!); Burton (Curtis 1930); Marchington (1948!); Stafford (1961!); Brereton; Hednesford.

M. officinalis (L.) Pall. Ribbed Melilot
Waste ground: locally abundant: L–M, (N), Y–b, e–f, j–k, n–t, w, y, 36. Brickyard, Codsall (Fraser 1865!).

M. alba Medic. White Melilot
Waste ground, rubbish dumps, slag heaps: L, (N), b, (f), (r), s, w, y, 9. Wightwick Wharf (Fraser 1873!).

M. indica (L.) All. Small Melilot
Casual: rubbish dumps and railway sidings: G, L, (N), b, (e–g), (r), 3. Wightwick (Fraser 1873!).

Trifolium pratense L. Red Clover
Grassland: A–z, 791. Yoxall Lodge (Gisborne 1791!).

T. medium L. Zigzag Clover
Hedgebanks and hayfields: B–C, E–P, R–U, W–g, j–y, 243.

'Common in hedges and ditch banks on the clayey soil in the parish of Blymhill' (Pitt 1794).

T. incarnatum L.   Crimson Clover
Relic of cultivation: E, (N), R, (S), (b), (r), s, (w), 3. Garner, 1844.

T. arvense L.   Haresfoot Clover
Sandy roadsides and gravel pits: K, (N), (R), W, (Y), (b), e, g, (h), m, r, w, 13. Dickenson, 1798. Frequent on Cannock Chase and about Wombourn and Kinver: elsewhere recorded recently for Betley Common (1941!); roadside near Loynton Moss (1954!); gravel pit near Wychnor (1956!); Lichfield.

T. striatum L.   Knotted Clover
Dry open habitats on sandstone and limestone: H–J, P, X, (Y), (b), (f), r, 7. 'In a field near Stafford' (Dickenson 1798). Recent records for roadside near Great Bridgeford station (1944!); Hall Dale and Wetton Mill (1947!); Sugarloaf near Wetton Mill (1948!); Bunster Hill (1951!); sand pit at Smestow (1952!); raddlepits near Wootton (1958!); Wolfscote Dale at the footbridge, SK/130584.

T. hybridum L.   Alsike Clover
Fields and roadsides: A–C, E–w, y, 323. Near Enville (Fraser 1865!).

T. repens L.   White Clover
Grassland: A–z, 798. Pitt, 1794.

T. campestre Schreb.   Hop Trefoil
Sandy fields and roadsides, gravel pits, limestone quarries, railway tracks and waste ground in old mining areas: H–M, (N), Q–U, W–Z, b, d–g, j–s, u, w–x, 80. Pitt, 1794. (Map p 209).

T. dubium Sibth.   Lesser Trefoil
Pastures and roadsides: B–C, E–z, 514. Brown, 1863.

T. micranthum Viv.   Slender Trefoil
Doubtful: small forms of *T. dubium* have been mistaken for it. But Purchas (1889) claimed to have seen it on the limestone.

Anthyllis vulneraria L.   Kidney-vetch
Dry banks in shallow, usually calcareous, soils: H–J, N–P, (S), (f), k, m, s–t, w, 14. Goscote near Walsall (Pitt 1796) and Barr Beacon (Pitt 1813). Recorded for Dovedale, the Manifold Valley and the Weaver Hills, for a cornfield at Swynnerton (Ridge 1916!) and a

meadow by the Trent near Armitage (Reader 1917!), and for slag heaps, railways and sand quarries in the south.

Lotus corniculatus L. Common Birdsfoot-trefoil
Grassland and waste places: A–z, 695. Pitt, 1794.

L. tenuis Waldst. & Kit. ex Willd. Narrow-leaved Birdsfoot-trefoil
Field near Brereton Cross (Reader 1922!).

L. uliginosus Schkuhr Greater Birdsfoot-trefoil
Marshy places: A–z, 535. Garner, 1844.

Galega officinalis L. Goat's-rue
Froghall, railway hedge not far from cottages (Allen 1910!); Heathfield Lane pools, Darlaston, SO/968965, F. M. Slater (1970).

Astragalus glycyphyllos L. Wild Liquorice
Bushy places chiefly in the neighbourhood of Stafford, but seen only once in recent years. 'By the roadside on the bank nearly opposite the Roman Catholic chapel at Stafford in the Friars' (Forster 1796); field between King's Bromley Hall and the Trent (Shaw 1801); lane from Tixall to Stafford, just beyond Kingston Pool (Clifford 1817); in bushes at top of Coton Field (Garner 1844); Aqualate, both sides (Fraser 1864); Acton near Stafford (Moore 1889!); under hawthorns on a bank between Seighford and Creswell overlooking the river (1958!).

A. odoratus Lam.
Has persisted at Burton for many years. First recorded under this name by Druce (BEC 1930 Rep) and seen most recently by M. E. Smith, SK/254252 (1970).

Ornithopus perpusillus L. Birdsfoot
Gravel pits and sandy roadsides: frequent: (B), E, G, K–M, (N), Q–R, (S), W, Y, (Z–b), e–g, j, m–n, r, (t), w, 35. Near Lichfield (Whately in Withering 1787); Woodmill gravel pit (Gisborne 1792!), (Map p 209).

Hippocrepis comosa L. Horseshoe-vetch
'On a rock, up the dell, behind the house in Wetton Valley' (Garner 1844). The rock was Sugarloaf, where the plant was rediscovered in 1946 and where it continues to thrive.

Onobrychis viciifolia Scop. Sainfoin
Casual or relic of cultivation: seen recently only at West Bromwich,

two plants on a slag heap at Great Bridge (1950): (H–J), (N), (b), (j), s, 1. Pitt, 1794.

Vicia hirsuta (L.) Gray   Hairy Tare
A common weed of cultivation in the centre and south: (F), G, K–L, (N), P–w, y–z, 207. 'Cornfields and meadows, in a light soil, the most pernicious of all weeds to corn' (Dickenson 1798). (Map p 209).

V. tetrasperma (L.) Schreb.   Smooth Tare
Cornfields, roadsides, railway embankments and waste ground across the centre and down the west side of the county: (N), S, U, W–a, (b), c–d, (f), g, j, r, v–w, 23. Dickenson, 1798.

V. cracca L.   Tufted Vetch
Hayfields, hedgerows, wet meadows, wood borders: A–z, 664. 'Vicia Cracca . . . has been observed in some parts to do so well in Meddows, that it advances all starven weak Cattle above anything yet known' (Plot 1686, p. 204).

V. sylvatica L.   Wood Vetch
Less common than formerly and now perhaps restricted to woods on the Carboniferous limestone. 'By the side of the water which supplies the mills at Stone adjoining the Radfords' (Forster 1796) and 'very plentifully in a wood . . . not far from Oulton' (Forster in Clifford 1817). Other unconfirmed records for Arley, Madeley, Moddershall, Tatenhill, Tamworth and Tittensor. Garner (1844) said it was most abundant on the limestone, where it coloured the hillsides by the profusion of its beautiful blossoms. It occurs in several places near Wetton Mill, at the entrance to Butterton tunnel and in a rocky defile below Deepdale Farm. Recorded also for Ecton Tor (Bloxam 1853); near Coldwall Bridge (Smith 1871); Wootton Woods (Goodall 1882); Froghall (Berrisford 1912!); Barrowhill (1952!); Big Peg's Wood (1965); Rushley Wood (1965).

V. sepium L.   Bush Vetch
Hedgebanks: A–z, 699. Dickenson, 1798.

V. lutea L.   Yellow-vetch
Casual. Burton (Nowers 1907!); Needwood Forest (BEC 1920 Rep); Wrottesley, at Perton cross roads (1954).

V. sativa L. Common Vetch
Common as a relic of cultivation but not systematically recorded.
Dickenson, 1798.

V. angustifolia L. Narrow-leaved Vetch
Gravel pits and sandy roadsides: E, G–H, K–w, y–z, 250. Yoxall
Lodge (Gisborne 1787!).

V. lathyroides L. Spring Vetch
Small forms of *V. angustifolia* were sometimes mistaken for this
species, but Dr. D. W. Shimwell tells me that a specimen collected
by Fraser from Kinver Edge in 1868 is correct.

Lathyrus aphaca L. Yellow Vetchling
Casual. Whittington Heath (Fraser 1864!); Burton (Nowers 1892!);
meadow near Dilhorne (1920); chicken run, Bar Hill, Madeley
(1926).

L. nissolia L. Grass Vetchling
Casual. Coton Field (Withering 1787); Belmont (Pitt 1794); near
Barton (Garner 1844 and Brown 1863); hen run, Oakamoor
(Berrisford 1931!).

L. pratensis L. Meadow Vetchling
Hayfields and waysides: A–z, 751. Pitt, 1794.

L. sylvestris L. Narrow-leaved Everlasting-pea
'At the Red Hill near Stone Park . . . the only place in which I have
ever found it' (Forster 1796).

L. montanus Bernh. Bitter Vetchling
Common in heathy pastures in the northern half of the county,
infrequent in the south: B–D, F–P, S–U, W–a, g, k, (m), r, v, y–z,
126. 'Woods in the Moorlands' (Bourne in Dickenson 1798).
(Map p 210).

ROSACEAE

Filipendula vulgaris Moench Dropwort
Weaver Hills (Gisborne 1797); hill pastures at Gradbach (Garner
1844); limestone pastures, Dovedale (Brown 1863); railway
embankment, Armitage (Moore 1897); field near Barr Farm
(Bagnall 1901); Branston salt district (Nowers 1907!); meadow
near Armitage (Reader 1919!); Bunster Hill (1952); waste ground
at Brereton colliery (1962!).

F. ulmaria (L.) Maxim.   Meadowsweet
Roadside ditches, canal and river sides and marshy places in
woods and meadows: A–z, 643. Pitt, 1794.

Rubus chamaemorus L.   Cloudberry
Oliver Hill, within a few feet of the summit, G. A. Lovenbury in
1941 (1955).

R. saxatilis L.   Stone Bramble
'Dovedale in rough rocky places' (Smith 1871); Apes Tor (Purchas
in Bagnall 1901); Hall Dale (1948).

R. idaeus L.   Raspberry
Woods and hedgebanks: B–z, 481. Pitt, 1794. With white fruit at
Clayton (Garner 1844).

R. caesius L.   Dewberry
Thickets and hedgerows in calcareous soils: rare. Wooded slope
above the river near Thor's Cave (1952!) and in other places in the
Manifold Valley; by the railway at Rolleston (1956); Whittington
near Lichfield, in the lane north of the golf course (1956!). Many of
the old records probably refer to hybrids with taxa of the *Triviales*
section.

R. fruticosus L. sensu lato   Bramble
More than 100 brambles have been recognised in Staffordshire as
distinct taxa, though only about two thirds of these can be named.
Some of the unnamed brambles are locally abundant and no doubt
deserve names, but the modern batologist must hesitate to increase
the already bewilderingly long list of brambles merely to improve
the tally for his county. His wisest course is probably to confine his
attention to taxa which are known to occur in other counties
besides his own, as A. Newton (1971) has recently done for Cheshire.
   A preliminary account of the Staffordshire brambles was published
in 1955 and a detailed account with a full range of maps is a work
for the future. All we can do here is to outline the distribution of
those brambles which already possess names. With a few exceptions
the nomenclature of W. C. R. Watson's *Handbook* (1958) has been
retained, so that readers will know where to find a description,
but the words 'sensu Watson' are used after names which may have
to be corrected later on. The records rest entirely on recent observations in the field and for that reason first records are not

distinguished. Specimens of every taxon are preserved in the author's herbarium.

An analysis of the sixty-four brambles in the following list yields some interesting results. Only fourteen can be described as common and widespread and even some of these are more likely to be found in some parts of Staffordshire than in others. The two commonest brambles, with the possible exception of *R. dumetorum* agg, are *R. vestitus* and *R. lindleianus*. Many are very local, though plentiful over a limited area, such as *R. sciocharis, R. obesifolius, R. daltrii, R. bloxamii* and *R. bellardii*. About a quarter of the total number have been found in only one or two places. These are outliers of large populations in other parts of the country and include *R. gratus, R. pyramidalis, R. silurum, R. rubritinctus* and *R. murrayi*.

R. scissus W. C. R. Wats.
Heaths and moors: locally common on sand or peat in exposed or shaded habitats, ascending to well over 1,000ft near Ramshaw Rocks: B–C, F–J, M–P, R, U, W, Z, e–g, j–n, r, t, w, 80. (Map p 210).

R. opacus Focke ex Bertram (*R. bertramii* G. Braun ex Focke)
Black Bank near Newcastle (1950!); Craddocks Moss (1951!); west side of Hand Leasow Wood near Uttoxeter (1953!); Shaw Wood (1953!).

R. plicatus Weihe and Nees
In two places near Cheddleton, viz Cats Edge, SJ/949523 (1966!), and at the foot of Hills Wood near the railway SJ/9950 (1967!).

R. accrescens A. Newton, *Watsonia* 8: 369 (1971).
Hanchurch Hills and other open heathy places in the north: locally abundant: B–C, F–G, K–N, R–S, 22.

R. eboracensis W. C. R. Wats.
Hedgerows: widespread and much commoner than the following records indicate: C, H, L, S, X–a, f–g, n, 14.

R. sublustris Lees
Common in hedgerows in the centre and south, but rare in the north, except on the limestone: H–J, L, N–P, R–s, u–v, x, 150.

R. balfourianus Bloxam ex Bab.
Barbers Gorse near Sheriff Hales, SJ/765145 (1961!).

R. warrenii Sudre
Common in many places on the Bunter sandstone near Whitmore, but apparently very local: H, L, S, 5.

R. myriacanthus Focke (sensu Watson)
Probably common in the north west, but under-recorded: K, N, R, 4.

R. dumetorum agg.
This is a convenient provisional name for a strongly armed, glandular and constantly white-flowered bramble of the *Triviales* section. It grows in hedgerows and waste places on low ground, often in damp situations, and is one of the commonest brambles in Staffordshire as in Cheshire. Records have not been kept.

R. gratus Focke
Bridle road through Bishop's Wood, SJ/7530 (1965!).

R. calvatus Lees ex Bloxam
Thickets and roadsides, particularly on sandstone: frequent: B, G, K–L, N–P, R–U, Y, e–f, j, 29.

R. sciocharis Sudre
Very local, but common in the lanes near Codsall, particularly County Lane (1953! 1965!), and on Kinver Edge (1951!).

R. carpinifolius Weihe and Nees
Sandstone heaths with Cannock Chase as its stronghold: C, G–H, K, N, R, T, W, Z–a, c–f, j–n, r, w, 63.

R. selmeri Lindeb. (*R. nemoralis* sensu Watson)
Another species of the sandstone with Cannock Chase and the heathy country round Lichfield and Whitmore as its chief centres: B, E, G, K–N, Q–S, W–a, c–g, j–n, q, t, 112. (Map p 210).

R. laciniatus Willd.
Groundslow Fields near Tittensor (1951!) and Farewell near Lichfield (1950!).

R. lindleianus Lees
Hedgerows and wood borders throughout the county, except in the north east: B, F–G, J–t, w, 335. (Map p 210).

R. egregius Focke (sensu Watson)
Sutton Park, the western edge between the Roman road and the modern road, SP/085975 (1951!).

R. robii (W. C. R. Wats.), A. Newton (*R. muenteri* sensu Watson)
Hedgerows in the north west: locally plentiful, especially in the neighbourhood of Ashley, Maer, Whitmore and Madeley: B–C, F, K–L, R–S, 11.

R. macrophyllus Weihe and Nees
Tall hedgerows and wood borders on sandstone: local: Rock Lane, Mucklestone (1951!); Hey Sprink (1952!) and lane to five pits, SJ/7643 (1953!), Madeley; Whitmore Wood (1962!).

R. amplificatus Lees
Scattered bushes in hedgerows on the western border of the county: elsewhere recorded at present only for a disused gravel pit at Brocton, SJ/967187 (1966!), and the roadside between Farewell and Lichfield (1951!): K, R, W–X, c–f, 10.

R. pyramidalis Kalt.
Tamworth, about the flag pole on the eastern edge of Hopwas Wood (1962!); Dog Lane, Hanchurch Hills, SJ/833392 (1967!).

R. obesifolius W. C. R. Wats.
Roadsides and bushy places near Cheddleton and Leek: local: B, G–H, 5. The type specimen was gathered from the roadside between Belmont and Cheddleton in 1950. There is a fine bush at the entrance to Coombes Valley near the barn, SK/005529 (1960!), and other bushes at Meerbrook (1946!).

R. incurvatus Bab.
Hedgerows and a field corner east and west of Blacklake Farm near Fulford (1960!); plentiful by the water below Consall Old Hall and by the roadside north of Blakeley (1967!).

R. favonii W. C. R. Wats.
Seckley Wood (1954!).

R. polyanthemus Lindeb.
Hedgerows, disused gravel pits and heaths throughout the county: B, F–G, K, M–N, Q–T, V–Y, a, d–g, j–k, n, r, w, 78.

R. rubritinctus W. C. R. Wats.
Clifton Campville, SK/267106 (1962!).

R. cardiophyllus Muell. and Lefèv.
Hedgerows on clay or sand: rare in the north, frequent in the south, though there are seldom many bushes in any one place: E, K–L, R–S, U, W–X, a, d–e, g, j–n, r, w, z, 34.

R. rotundatus P. J. Muell. ex Genev.
Rough heathy field at Offleybrook near Eccleshall, SJ/785305 (1960!).

R. lindebergii P. J. Muell.
Bushy hillsides in the north: locally plentiful: B–D, G–H, P, U, 16.

R. silurum (A. Ley) W. C. R. Wats.
Lane at Blorepipe near Eccleshall, SJ/768305 (1963!); roadside west of Wood Corner, SJ/750297 (1965!).

R. ulmifolius Schott
A lowland bramble abundant on the Keuper marl of central and southern Staffordshire, less common on sandstone and apparently quite absent from large areas of the north, though it occurs on limestone in the Manifold Falley: E, H–K, P, S–w, z, 211. (Map p 211).

R. pseudobifrons Sudre (sensu Watson)
Whittington near Lichfield, in the lane north of the golf course, SK/146080 (1966!).

R. procerus P. J. Muell.
Of garden origin: formerly on Whitmore Common (1950!); luxuriantly in an old marl pit, now a dump, at Outwoods, SK/223254 (1961!).

R. falcatus Kalt.
Haunton, in the lane to the river (1946!).

R. sprengelii Weihe
Moist woods and heathy banks, chiefly in the north: common about Leek and in the Dane valley: A–H, K–P, R–U, X, Z–a, e, g, j–m, t, 132. (Map p 211).

R. arrhenii (Lange) Lange var *polyadenes* Gravet ap. Focke
Roadside at the foot of Bailey's Hill near Biddulph, SJ/8958 (1951!).

R. lasiostachys Muell. and Lefèv. (sensu Watson)
Pound Green Common near Upper Arley (1959!).

R. vestitus Weihe and Nees
One of the commonest Staffordshire brambles on limestone, sandstone and Keuper marl: A, C, E–n, q–x, 354. In the Manifold Valley its flowers seem to be always white, but elsewhere bushes

with showy red flowers are frequent, sometimes, as at Ashley, growing side by side with white-flowered plants. (Map p 211).

R. criniger (E. F. Linton) Rogers
Common in central Staffordshire, especially on the west side of Lichfield: K–P, S–U, X–a, c–g, j, m–n, t, 84.

R. eifeliensis Wirtg. (sensu Watson)
Locally plentiful in the north near Biddulph, Rushton and Leek: B, F–G, L, 5.

R. mucronulatus Bor.
On the west side of Hand Leasow Wood near Uttoxeter (1953!) and around the flag pole on the east side of Hopwas Wood near Tamworth, SK/179057 (1966!).

R. daltrii Edees and Rilstone
Wood borders west of a line from Newcastle to Eccleshall: plentiful in this restricted area: K–L, R, 16. The type specimen was gathered in 1945 from a triangle of rough shaded ground north of Whitmore Common, SJ/799422.

R. taeniarum Lindeb.
Common about Maer and Mucklestone and extending south to Penkridge: R–S, X, e, 8.

R. echinatus Lindl. (*R. discerptus* P. J. Muell.)
Hedgerows: frequent in the south, rare in the north: K, S–b, d–f, h–n, q–r, w, 42.

R. echinatoides (Rogers) Sudre
Frequent in the north, absent from the south: B, F–H, K–P, R–S, 22.

R. granulatus Muell. and Lefev. (sensu Watson)
Park Lane, Whittington, SK/165098 (1962!).

R. rubristylus W. C. R. Wats.
A frequent bramble of sandy soil in the west and on Cannock Chase: K–L, Q–T, W–Y, c–f, j–m, r, 59.

R. adenanthoides A. Newton, *Watsonia* 8: 374 (1971).
In a damp copse at Whitmore (1951!); roadside bank between Heleigh Castle and Cooksgate (1951!); woodland by the river at Consall Forge (1950!).

R. bloxamii Lees
Plentiful between Lichfield and Tamworth, but almost confined to this area: f–g, n–p, t, 13.

R. pallidus Weihe and Nees
Woods and shady hedgebanks: local: N, R–S, U, W–X, Z, 9. One of the commonest brambles in the Gnosall and Norbury district, being abundant round Loynton Moss and in Shelmore Wood, Mill Haft, Coneygreave Haft and elsewhere. It can also be seen near the water tower on Hanchurch Hills and at Consall Forge.

R. distractiformis A. Newton, *Watsonia* 8: 375 (1971).
Locally common near the Cheshire border in the north of the county. West side of Rudyard reservoir (1949!); Meerbrook, lane behind the church (1960!).

R. euryanthemus W. C. R. Wats.
Sometimes rampant in wet woods: recorded for many places north and west of Newcastle, infrequent elsewhere: E–F, K–L, R–S, W–X, e, m, 22. Abundant under the trees round Betley Mere.

R. insectifolius Muell. and Lefèv.
Disused gravel pit at Bloreheath, SJ/712353 (1950!); wood at Napley Heath near Mucklestone, SJ/7138 (1954!); lane from Hopwas to Dunstall Bridge, SK/1804.

R. scaber Weihe and Nees
Very local. Hanch Wood at the railway bridge (1966!) and along Tuppenhurst Lane near King's Bromley.

R. rufescens Muell. and Lefèv.
A woodland bramble: local in the south. Seckley Wood (1954!); Kingswood Common (1954!); woods near Barton (1962!) and Lichfield (1962!); Baggeridge Wood (1963!); Knightley Gorse covert near Gnosall (1968!).

R. pascuorum W. C. R. Wats.
Kinver Edge (1952!); sandy roadsides at Seisdon (1953!); Hollies Common near Gnosall (1964!).

R. leightonii Lees ex Leighton
Hedgerows in the south east: abundant near Lichfield, infrequent elsewhere: K, N, R–S, Y–a, e–r, t, w, 72. (Map p 211).

R. diversus W. C. R. Wats.
Below the mill between Farewell and Lichfield, SK/087115 (1951!),
but now possibly extinct.

R. murrayi Sudre
Whitmore, SJ/799422 (1950!), a single bush; roadside by wood
between Swindon and Highgate Common, SO/8490 (1953!).

R. hylocharis W. C. R. Wats.
Frequent in woods: B, F–H, K–M, P–U, W, Y, c, f, j, m–n, 30.

R. dasyphyllus (Rogers) E. S. Marshall
The commonest glandular bramble, except *R. dumetorum* agg:
B–C, F–H, K–U, W–b, d–g, j–t, w, 197. (Map p 212).

R. bellardii Weihe and Nees
Plentiful in the Churnet Valley woods, where it can be seen in
Hawksmoor Wood, Mud-dale Wood (1952!) and along the whole
length of the Earl's Drive. Elsewhere in the county it is recorded
for Milford covert, SJ/969209 (1966!), and Hillfields Coppice near
Upper Arley, SO/785819.

R. lintonii Focke ex Bab.
Whitmore, between the village and the Hall, SJ/809411 (1951!),
and in the lane on the north side of Whitmore Common (1952!).

Potentilla palustris (L.) Scop. Marsh Cinquefoil
Peat swamps, lake margins and pits in fields, often growing in the
water: local: B–C, E–H, K–M, (N), R–T, W–Z, (a), (c), d–g, j,
m–n, (r), t, w, 49. 'On the bog at Willowbridge' (Waring 1770).
'In the pit in the Foxholes where the Buckbean grows, 1792'
(Gisborne MS). (Map p 212).

P. sterilis (L.) Garcke Barren Strawberry
Hedgebanks: common, except in very acid soils: B–g, i–k, n, q–s,
v–w, z, 315. Pitt, 1796.

P. anserina L. Silverweed
'Roadsides and pastures in a wet soil' (Dickenson 1798): in both
dry and damp sandy and gravelly places and on the stony shores of
meres and reservoirs: A–C, E–x, z, 551. Pitt, 1794.

P. argentea L. Hoary Cinquefoil
Bank near King's Bromley (Power 1823!); Kinver, 1862 (Mathews
1884); Trescott (Fraser 1865!); Whittington Heath (Fraser 1877!);

wall tops, Armitage (Reader in Bagnall 1901 and Reader 1917!);
Forton, at the top of the field above Windswell Pool (1954!);
footpath on the western slope of Kinver Edge (1968).

P. recta L.   Sulphur Cinquefoil
Hoar Cross (Miss Meynell in BEC 1930 Rep); near Upper Arley,
SO/757790 (1959!).

P. norvegica L.   Ternate-leaved Cinquefoil
Casual. Oakamoor (Berrisford 1903!); Burton (BEC 1927 Rep);
North Street, Stoke (1933!).

P. tabernaemontani Aschers.   Spring Cinquefoil
Confined to the Carboniferous limestone, where it grows in a few
places over the rocks which protrude from the hillsides, particularly
near Wetton Mill. There used to be a plant on the highest point of
Thor's Cave hill. Garner (1867) saw it in Dovedale, probably in
1832, but we do not know in which county. Other old records for
Wetton (Carrington in Smith 1871) and Ecton Hill (Worsley in
Watson 1874). Recent records for a field near Sparrowlee (1943!);
Ecton Hill (1943!); Wetton, in several places (1943); rock at
Weag's Bridge (1943!); Hall Dale (1948); Ilam, on a flat rock near
Castern overlooking the valley (1948).

P. erecta (L.) Räusch.   Tormentil
'In woods and moist old pastures, common' (Pitt 1796); 'Common
in heathy places' (Garner 1844): A–z, 482. When Babington was
staying at Yoxall Lodge in 1832 he wrote in his diary: 'Sept 1.
Gathered 759 specimens of *Potentilla Tormentilla* to examine . . .'
(A.M.B. 1897). Two days later he gathered 1,632 specimens.

P. anglica Laichard.   Trailing Tormentil
'Rather local in its distribution, but occurs in many places in woods
and hedgebanks' (Ridge in Flora). This is doubtless true, but
because of the difficulty of distinguishing this species from hybrids
with *P. erecta* and *P. reptans* and from *P. erecta* x *reptans* field
records have not been kept. Biddulph (Painter in BEC 1887 Rep).
Specimens in my herbarium from Audley, Balterley, Barlaston,
Dilhorne, Eccleshall, Hednesford, Ipstones, Maer, Milwich and
Ramshorn. Marshall (1893) quotes records of *P. anglica* x *reptans*
for Ilam Moor and Dovedale, and of *P. anglica* x *erecta* for Rocester,
Rudyard and near Longnor.

P. reptans L. Creeping Cinquefoil
Hedgebanks, railway embankments and roadsides: A–B, E–z, 513.
Pitt, 1794.

Fragaria vesca L. Wild Strawberry
Hedgebanks, railway embankments, dry wood clearings and rubble,
especially in base-rich soils: B–C, F–U, W–g, i–n, r–s, v–w, z, 228.
Pitt, 1794. (Map p 212).

Geum urbanum L. Wood Avens
Woods and shady hedgerows: B–C, E, (F), G–k, n–w, z, 453.
Pitt, 1794. (Map p 212).

G. rivale L. Water Avens
Damp pastures and roadsides and by streams and runnels in woods
and upland hayfields, especially on the Carboniferous limestone:
plentiful in the north east, infrequent elsewhere: C–D, F–P, R–U,
W–X, (Y), (b), c, (d), (m–n), (r), 75. Near Wootton (Gisborne
1793!). 'One of the most elegant wild plants of this island, grows
about Ilam' (Gisborne in Pitt 1796). *G. rivale* x *urbanum* is often
seen where the parents grow together. (Map p 213).

Dryas octopetala L. Mountain Avens
Druce (1932) quotes Staffordshire for this species, but the only
evidence I have found is that supplied by Garner (1844): 'Mow
Cop, on the authority of the late Dr Davidson. Several gardeners
show the plant as obtained there.' Garner himself never found it,
though he says (1878) he looked for it more than once. But he
thought the habitat, 'barren grit hills and rocks,' suitable. It is,
however, quite unsuited to a lime-loving species.

Agrimonia eupatoria L. Agrimony
Hedgebanks, particularly on the Keuper marl: H–L, (N), P–k,
n–t, v–w, y–z, 165. Pitt, 1794. (Map p 213).

A. procera Wallr. Fragrant Agrimony
Tixall, old marl pit (Bagnall 1896!); roadside between Wetton and
Wetton Mill (1941!); Brewood, lane to the Hattons (1954!).

Alchemilla vestita (Buser) Raunk. Lady's-mantle
Meadows: the commonest species of the genus in Staffordshire
and the only one recorded for the south: B, G–e, (f), g–j, r, t, v–w,
y–z, 153. Pitt, 1794, as *A. vulgaris* L. (Map p 213).

A. xanthochlora Rothm.

Hill pastures: common in the north: B–C, G–J, M–N, S, 52. Oakamoor (Berrisford 1902!). (Map p 213).

A. glabra Neygenf.

Hill pastures: common in the north: B–D, G–J, M–P, 65. Morridge near Leek (T. E. Routh in WBEC 1908–9 Rep). (Map p 214).

Aphanes arvensis L.   Parsley-piert

Dry shallow soil on sandstone and limestone: common about outcrops of rock on limestone hillsides and in sandy fields and on heaths in the south: H–L, N–P, R–S, W–Y, b, d–j, n, r, v–w, 82. Near Yoxall Lodge (Gisborne 1792!).

A. microcarpa (Boiss. and Reut.) Rothm.

Overlooked until recently and probably commoner in acid soils than the following records imply: Kinver Edge (1950!); Swindon (1952!); Cranmoor Wood (1954!); Sandon (1956!); Offleybrook (1960!).

Sanguisorba officinalis L.   Great Burnet

Damp hayfields and railway embankments: common throughout the county except on the western edge: B–E, (F), G–J, (K), L–P, S–b, d–h, k–u, w, y, 237. 'In Yoxall Lodge meadow' (Gisborne 1791!). (Map p 214).

Poterium sanguisorba L.   Salad Burnet

Abundant on the limestone but almost confined to it: C, H–J, N–P, (b), (g), h, (j), m, (t), (x), 30. Pitt, 1794.

Rosa L.   Roses

Wild roses have attracted Staffordshire botanists from the earliest days and apart from a multitude of published records there are many specimens in the collections made by Fraser, Bagnall, Reader and others, which must be revised before a complete account of the genus can be written. For the present all the old records and specimens of Dog Rose and Downy Rose have been ignored. There are two recent sources of information, a study of the roses of the Leek district made by J. Thompson and Prof J. W. Heslop-Harrison (1955) and a personal collection of ninety Staffordshire specimens which Mrs H. R. H. Vaughan determined for me in 1966 and 1968. In the following account an exclamation mark after a recent date indicates a specimen in my herbarium determined by Mrs Vaughan. All undated records were contributed by J. Thompson or Prof Heslop-Harrison. As both J. Thompson (with Prof Heslop-Harrison) and Mrs Vaughan followed the classification of Wolley-Dod (1931), that arrangement has been retained here. But to save space species have not been subdivided further than varieties. The division of varieties into forms has been ignored. Most of the records come from the north of the county and it may be thought that they give a false perspective. But in fact the north is richer than the south and, though there is much detail to be filled in, the general picture is probably true.

R. arvensis Huds.   Field Rose
Hedges: A–z, 529. 'Rough ground near Russell's Hall, where the coal has been got' (Wainwright in Shaw 1801). Var *vulgaris* Ser. is the typical plant and is common. Var *ovata* (Lej.) Desv.: New Inn, Longsdon; canal feeder, Leek; Deep Hayes, Cheddleton. Var *biserrata* Crep.: Middleton Green, Leigh (1964!).

R. pimpinellifolia L.   Burnet Rose
Unconfirmed records for Yoxall (Gisborne 1791!), 'In the lane from the Forest to Yoxall, abundantly' (Gisborne MS); Betley (Pitt 1817); Amblecote (Scott 1832).

R. canina L.   Dog Rose
Hedgerows throughout the county: B, E–P, R–k, n–s, u, w–y, 174. Pitt, 1794. Var *lutetiana* (Lem.) Baker, var *sphaerica* (Gren.) Dum., var *flexibilis* (Desegl.) Rouy, var *senticosa* (Ach.) Baker, var *spuria* (Pug.) W.-Dod, var *globularis* (Franch.) Dum., var *ramosissima* Rau, var *dumalis* (Bechst.) Dum., var *biserrata* (Mer.) Baker, var *fraxinoides* H.Br., var *schlimperti* Hofm. and var *sylvularum* (Rip.) Rouy are represented by specimens, and var *rhynchocarpa* (Rip.) Rouy, var *stenocarpa* (Desegl.) Rouy, var *andegavensis* (Bast.) Desp., var *blondaeana* (Rip.) Rouy and var *latebrosa* (Desegl.) N.E.Br. are recorded.

R. dumetorum Thuill.   Dog Rose
Var *typica* W.-Dod: Bradnop, railway embankment (1965!); Apes Tor (1964!); Rolleston, Dove Bridge (1964!); field below Thor's Cave (1964!); Abbots Bromley (1966!); Sandon Bank (1966!); Longdon Green (1966!); Bradley (1964!); Leigh, roadside between Bearsbrook and Godstone (1964!); Deep Hayes; Longsdon, canal bank. Var *ramealis* (Pug.) W.-Dod: Head of

Rudyard Lake. Var *gabrielis* (F.Ger.) R.Kell.: Longdon Green
(1966!). Var *calophylla* Rouy: Whitmore (1946!). Var *sphaerocarpa*
(Pug.) W.-Dod: Butterton near Whitmore (1943!); field below
Thor's Cave (1964!); Newcastle (1943!); Heathylee (1947!). Var
erecta W.-Dod: Weston upon Trent, path by church to road
(1964!).

R. afzeliana Fr.   Dog Rose
Var *glaucophylla* (Winch) W.-Dod: Blore with Swinscoe (1947!);
Hollinsclough, near Dun Cow's Grove (1947!); Meerbrook
(1950!); Washgate; Longsdon, canal bank; Leek, Pickwood Dene;
Cheddleton; Highshutt; Rushton; Cauldon. Var *reuteri* (God.)
W.-Dod: Washgate; near Crowdecote (Staffs); Winkhill; Highshutt.
Var *stephanocarpa* (Desegl. et Rip.) W.-Dod: Longsdon, canal
bank; head of Rudyard Lake. Var *oenensis* (R.Kell.) W.-Dod:
Washgate. Var *subcanina* (Chr.) W.-Dod: Leek, canal bank; head
of Rudyard Lake. Var *denticulata* (R.Kell.) W.-Dod: Washgate.

R. coriifolia Fr.   Dog Rose
Var *typica* Chr.: Whitmore (1943!); Waterhouses, in the Hamps
valley (1947!); Washgate; Wall Grange; Longsdon; Rushton;
Longnor; Warslow, Clough Head; Cheddleton; Leek, Pickwood
Dene. Var *watsonii* (Baker) W.-Dod: Rudyard, canal feeder. Var
bovernieriana Chr.: Leek, canal bank. Var *bakeri* (Desegl.) W.-Dod:
Rushton; Washgate.

R. obtusifolia Desv.   Dog Rose
Var *typica* W.-Dod: Longsdon, New Inn. Var *sclerophylla* (Scheutz)
W.-Dod: Heaton. Var *borreri* (Woods) W.-Dod: Waterhouses, in
the Hamps valley (1947!).

R. sherardii Davies   Downy Rose
Var *typica* W.-Dod: Heathylee (1952!); Seven Acre Wood near
Cheadle (1958!); field below Thor's Cave (1964!); Kingsley,
roadside at the entrance gate to Shaw Hall (1958!); Farley (1958!);
Bradley (1964!); Washgate; Cheddleton; Leek, Pickwood Dene;
Longsdon, New Inn. Var *omissa* (Desegl.) W.-Dod: Apes Tor
(1964!); near Warslow on road to Onecote (1964!); High Offley
(1948!); Fawfieldhead, Piggenhole (1964!); Horton (1964!). Var
woodsiana W.-Dod: Ipstones Edge, south of Upper Cadlow
(1965!). Var *suberecta* (Ley) W.-Dod: field below Thor's Cave
(1964!); Morridge; Washgate; Longsdon, canal bank; Leek. Var

cinerascens (Dum.) W.-Dod: Wetton, field below Thor's Cave (1964!).

R. villosa L.   Downy Rose
Var *mollis* Sm.: Washgate; Ipstones Edge; Leekbrook, railway side; Hollinsclough; Warslow, Clough Head; head of Rudyard Lake.

R. rubiginosa L.   Sweet Briar
Hedges and thickets: no recent records. 'Between Dudley and Tipton' (Stokes in Withering 1787); hedges near Belmont (Sneyd in Dickenson 1798); Darlaston, Willoughbridge, Hill Chorlton, Maer, Whitmore (Garner 1844); Anslow, 'but generally planted' (Brown 1863); 'In hedges occasionally' (Smith 1871); Uttoxeter (Nowers 1899!); 'Field road above Oakamoor' (Fraser 1888!); Cheadle (1912); 'Oaken, Staffs!, 1838, Dr Bidwell' (Druce in BEC 1918 Rep).

R. micrantha Borrer ex. Sm.   Sweet Briar
Hedges: rare. Lane from Hill Ridware to Rugeley, Mrs E. Reynolds, 1848, specimen seen by Reader (1923); near Teddesley (Bagnall 1897!). Var *hystrix* (Lem.) Baker: Rushton.

Prunus spinosa L.   Blackthorn
Hedgerows and scrub: B–z, 591. Pitt, 1794.

P. domestica L.   Wild Plum
Frequent in hedges, but usually near houses. Pitt (1796) said it was common in hedges, but Garner (1844) thought it rarely, if ever, wild.

P. avium (L.) L.   Wild Cherry
Woods and hedgerows, particularly in the west and centre of the county, usually as single trees: A, (B), E, G–U, W–k, n–r, t, v–x, z, 139. Stewponey (Fraser 1877!).

P. cerasus L.   Dwarf Cherry
'Frequently wild in woods' (Garner 1844), but there are few records and no recent ones. Near Denstone (Edwardes 1876); near Stone (Tylecote 1885); Yarlet, Hanchurch and near Rushton (Ridge in Flora).

P. padus L.   Bird Cherry
Copses and bushy places: frequent in the north, especially in

moorland valleys and on the Carboniferous limestone: B–D, F–T, (W), X, Z, k–n, (p), 60. 'Growing in a hedge on this farm (Pendeford); also on Bushbury Hall Farm, about Shelford near Walsall, and in the Moorlands' (Pitt 1794).

Cotoneaster microphyllus Wall. ex Lindl.   Small-leaved Cotoneaster
'Bird sown, in turf on limestone downs near Blore' (Druce in BEC 1926 Rep). Naturalised in a few places in this neighbourhood.

Crataegus laevigata (Poir.) DC.   Midland Hawthorn
Hedges: introduced: uncommon: L, (Y), (a–b), (g), (m), (v–w), 2. Hedge near Whittington Heath, Kinver (Fraser 1877!). Recent records for Seabridge Lane (1942!); Silverdale (1943!); Butterton Park (1944!).

C. monogyna Jacq.   Hawthorn
Woods, limestone hillsides and planted in hedges: A–z, 790. 'The common white hawthorn is well known everywhere' (Pitt 1794).

C. prunifolia (Poir.) Pers.
A few yards to the south of the military cemetery on Cannock Chase and on the opposite side of the road there is a thorn tree which has long been known as the Cank Thorn or Boundary Thorn. Wright (1934) tells us that it has been a landmark for 600 years. The present introduced species is doubtless the successor of an old native bush.

Sorbus aucuparia L.   Rowan
Often planted, but native in the north of the county, especially in the gritstone cloughs: A–g, i–y, 428. 'A hardy shrub, common in hedges, but flourishing well on rocky and bleak hills' (Pitt 1796). Dickenson (1798) who admired its 'elegant foliage and charming scarlet berries' said it was 'frequently seen growing spontaneously in a thin, meagre, wet soil, where scarcely any tree except this and the birch will flourish.'

S. aria (L.) Crantz sensu lato   Common White-beam
Frequently seen in parks and shrubberies, but as a native tree perhaps confined to the limestone. 'In Dovedale and elsewhere in the Moorlands' (Sneyd in Pitt 1796). 'On high limestone rocks; several trees on Beeston Tor; fine trees at Beresford; Mill Dale; abundant in Dovedale' (Garner 1844). Purchas (1885) said that many of the bushes in Dovedale were inaccessible, but that all

which he had been able to examine belonged to *S. rupicola* (Syme) Hedl.

S. torminalis (L.) Crantz   Wild Service-tree
Old records for Pendeford (Pitt 1794); Uttoxeter (Watson 1837); Trentham Park, where Garner (1844) tells us there were some very large and ancient trees over the brow of the hill towards Nowall; near Longton, near Upper Arley, about Rolleston (Garner 1844); Needwood Forest, Henhurst, Knightley Park (Brown 1863); Seckley Wood (Fraser 1865!). Recent records for Upper Arley and Seckley Wood (1954!); Abbots Bromley, roadside between Ash Brook and Hart's Farm (1959!); Bagot's Park, SK/086265, one fine tree remaining (1970).

Pyrus communis L.   Wild Pear
Single trees have been recorded occasionally for roadsides near houses. Dickenson, 1798.

Malus sylvestris Mill.   Crab Apple
Hedgerows, copses, hillside birch woods and by the sides of ponds: B–z, 434. Pitt, 1794. Subsp *sylvestris* and subsp *mitis* (Wallr.) Mansf. both occur. The former is the wild plant and is the commoner in the north of the county. Plot (1686, p 226) tells us that apples abound in the parish of Arley, 'where all the grounds and hedges are planted, much after the manner of Worcestershire (into which indeed it runs with a long nook), there being scarce a cottage that has not some proportionable plantation belonging to it, having all sorts of pippins of the best.'

CRASSULACEAE

Sedum telephium L.   Orpine
Frequent in stony places and on walls in the Manifold Valley and Dovedale, infrequent on hedgebanks elsewhere: G, H–J, N, (R), U, d, m, (n), (r), (x), 16. 'Pasture near Mr Pearson's house, Tettenhall' (Withering 1787). 'Growing on roofs in the Moorlands, particularly at Wetton' (Pitt 1794).

S. acre L.   Biting Stonecrop
Stony places on the limestone, less common elsewhere 'on walls, roofs and dry ground' (Brown 1863): C, G–J, (K), M–P, T, (Y), Z, b, e, (f–g), m, p, r–s, (x), 42. Pitt, 1794. (Map p 214).

S, reflexum L.   Reflexed Stonecrop
Burton Abbey walls (Dickenson 1798) and other old records.

Sempervivum tectorum L.   Houseleek
Old records for roofs and walls. Pitt, 1794.

Umbilicus rupestris (Salisb.) Dandy   Navelwort
Old stone walls: rare and decreasing. 'Much more common in . . .
Staffordshire . . . than more southward' (Waring 1770); rocks under
Heleigh Castle bank (Dickenson 1798); Kinver Edge (Scott 1832);
Dovedale, Ilam, and abundant on a bank between Endon and
Leek (Garner 1844); on a wall at Basford near Leek (1917 and
1951!); water trough at Sharpcliffe Hall (1939 and 1953). Garner
knew it 'in the dark fosse of Heyley Castle' in 1844. It was last
recorded for Endon in 1914.

SAXIFRAGACEAE

Saxifraga spathularis x *umbrosa*   London-pride
Naturalised in a few places. Belmont Woods, 1837, and abundantly in a rocky dell below Upper Cotton (Garner 1844), but not
seen in either place in 1897 (Bagnall 1901); ravine near Stockton
Brook (1912); ravine at Hollinsclough; Apes Tor (1964!).

S. tridactylites L.   Rue-leaved Saxifrage
Common on walls and about rocks in the limestone district:
H–J, N–P, (S), U, (Y), (b), (m), (s), (w), 23. Near Wootton
(Gisborne 1793!).

S. granulata L.   Meadow Saxifrage
Calcareous pastures and riverside meadows: local: C, (F), H–J,
N–P, U, Z, (b), (f), g–h, n–r, (t), w, 54. Pitt, 1794. (Map p 214).

S. hypnoides L.   Mossy Saxifrage
Confined to the Carboniferous limestone, but common there on
broken ground: H–J, 13. Dovedale (Gisborne 1793!).

Chrysosplenium oppositifolium L. Opposite-leaved Golden-saxifrage
Common in wet shady places, especially about springs and in
woods: A–U, W–c, e–f, i–n, q–r, v–w, 202. 'In the dingle at
Hanbury' (Gisborne 1792!). (Map p 215).

C. alternifolium L.   Alternate-leaved Golden-saxifrage
Wet shady places, but with a strong preference for limestone:
often growing with *C. oppositifolium* but rarer: (A), B, (G), H–J,

(K), N–P, T, a, e–f, (g), (i), m, (n), (r), w, 26. 'In shady woods near rills of water' (Withering 1776). Belmont (Sneyd in Pitt 1794). (Map p 215).

PARNASSIACEAE

Parnassia palustris L.   Grass-of-Parnassus
Damp hollows on the limestone and bogs and boggy meadows where the water is base-rich: local: H–J, N–P, (R–S), (W), (Z), d–e, (w), 9. Stokes in Withering, 1787. Ipstones Edge (1940!); Brocton (1945!); old quarries about Caldonlow and the Weaver Hills (1958!); north facing hillside near Thor's Cave (1964); Allimore Green Common (1967).

GROSSULARIACEAE

Ribes rubrum L.   Red Currant
Bird sown in hedges, but native in marshy woods: frequent: D–E, G–M, Q–U, W–a, (b), c–e, (f), g–h, k–r, w, y–z, 74.

R. nigrum L.   Black Currant
Hedges, riversides and wet places in woods: B, E–F, H, K–L, (M), N–P, R, W, Z–a, (b–c), d, f, (t), (w), 16. 'I have found this shrub in thickets and remote situations' (Pitt 1796).

R. alpinum L.   Mountain Currant
Limestone rocks: locally abundant in the Manifold Valley, as about Wetton, and in Dovedale: long known for Needwood Forest (Carter 1839), where, however, it may have been introduced (Brown 1863): H–J, a–b, 15. 'In a hedge not far from Ilam' (Sneyd in Withering 1796).

R. uva-crispa L.   Gooseberry
Hedges and woods: common, but usually bird sown: C–D, G–P, R–U, W–f, (g), j–r, t, v–w, y, 140. 'In hedges and have seen it on a church tower' (Pitt 1796).

DROSERACEAE

Drosera rotundifolia L.   Round-leaved Sundew
In sphagnum swamps on heaths and moors: once frequent, now rare or very local: (C), E, K, (M), N, (P), R–S, (Y), Z, (a), (c), e, (f–g), (m), 8. 'On the hillocks, called triddle-bogs, on the great bog at Willowbridge' (Waring 1770). In 1844 Garner said it was common and general in bogs. But today it is known only for

Balterley Heath, Craddocks Moss, the Downs Banks, the Ranger, Sherbrook Valley and Oldacre Valley on Cannock Chase, and Chartley Moss. These are all lowland stations. It seems to have disappeared from Goldsitch Moss, where Bloxam found it in 1853. (Map p 215).

D. anglica Huds.   Great Sundew
Bog in Oldacre Valley on Cannock Chase (Druce in BEC 1919 Rep). Still there in 1970.

D. intermedia Hayne   Oblong-leaved Sundew
Apparently extinct. 'Plentifully, but less so than *D. rotundifolia*, on the hillocks, called triddle-bogs, on the great bog at Willowbridge' (Waring 1770); Fairoak and Balterley (Garner 1844); Chartley Moss (Clifford 1817); Cannock Chase (Brown 1863), unless this is a mistake for *D. anglica* which Brown does not mention. Ridge (Flora) claimed to have seen it on Chartley Moss.

LYTHRACEAE

Lythrum salicaria L.   Purple Loosestrife
Marshy ground by streams and lakes in the centre and east of the county, particularly between Burton and Tamworth: E, S, W–Z b–e, (f), g–p, t, v, y, 46. Banks of Trent at Darlaston (Forster 1796) (Map p 215).

Peplis portula L.   Water Purslane
Wet muddy ground at the edges of ponds: widespread, but uncommon: B, (F), K, (T), U, (a), (e–f), g, k, (r), (t), (x), 6. Gravel pit beyond Woodmill Brook, 1791 (Gisborne MS). Recent records for dried-up bed of Rudyard reservoir (1934 and 1955!); King's Bromley, marshy hollow in a field between Rileyhill and the canal (1945!); Hatherton, margin of a dried-up pool between Heath Farm and Watling Street (1947!); Madeley (1951!); Leigh, Birchwood Park (1953!); Calf Heath Wood (1954!).

THYMELAEACEAE

Daphne mezereum L.   Mezereon
Unconfirmed records for Needwood Forest: 'In the dingle at Hanbury' (Gisborne 1792!) and near Byrkley Lodge (Brown 1863). Long known for Bincliff Thicket in the Manifold Valley (Carrington in Garner 1860), where a single bush was seen in 1967. Recently it has been found near Thor's Cave (Shimwell 1968), two old bushes

and four saplings 'in an inaccessible cliff community dominated by *Corylus*.'

D. laureola L.   Spurge-laurel
Native in Needwood Forest (Pitt 1794). Can still be found in Forest Banks in several places (1950!), in Banktop Wood, at Needwood House and in the old lane near Gilleon's Hall (1959!). Recently Shimwell (1968) has seen it in Cheshire Wood in the Manifold Valley.

ONAGRACEAE

Epilobium hirsutum L.   Great Willowherb
Stream sides, river banks, ditches, canals: A–z, 705. Pitt, 1794.

E. parviflorum Schreb.   Hoary Willowherb
Streams and diches, but particularly common by the canals: B–C, G–b, d–u, w–y, 281. Dickenson, 1798.

E. montanum L.   Broad-leaved Willowherb
Woods, hedgebanks, walls, gardens: B–z, 451. Dovedale (Gisborne 1792!). *E. montanum* x *obscurum*: Fawfieldhead, meadow near Ludburn Ford (1941!), det G. M. Ash.

E. roseum Schreb.   Pale Willowherb
Roadsides, walls, canal tow paths, gardens: frequent: B, G–J L–Q, (R), S–V, X–Y, a, (b), d–k, (m), n–t, w, y, 60. Pipe Marsh (Jackson 1837).

E. adenocaulon Hausskn.   American Willowherb
Stream sides, drains and wood clearings: probably frequent, but overlooked until recently: L, S, W, Y, b, g, n, s, y, 10. Burton (Brenan and Chapple in BEC 1937 Rep).

E. tetragonum L.   Square-stalked Willowherb
Uncertain. Bagnall claimed to have seen it near Fradley and at Aldridge (1901) and Ridge at Ashley (Flora).

E. obscurum Schreb.   Short-fruited Willowherb
Damp woods and marshy places throughout the county: B–D, F–r, u, w, 304. Burton (Brown 1863). *E. obscurum* x *parviflorum*: Barlaston Common (1941!), det. Ash. *E. obscurum* x *palustre*: Morridge moors at 1,500ft (1941!), det Ash.

E. palustre L.   Marsh Willowherb
Marshes: particularly common in peaty soil by moorland streams

and on Cannock Chase: B–D, G–K, M–N, R–Z, (a), e–f, (g), h, k–m, p, r, t, (v), (z), 128. Near Yoxall Lodge (Gisborne 1791!). Purchas (1885) noticed its absence from limestone. (Map p 216).

E. nerterioides Cunn.   New Zealand Willowherb
Trentham Park, several patches on moist gravel under young birch trees near Jacob's Ladder, R. H. Brown (1956!); Longton, stone and gravel pavement in the war memorial garden (1957); Wall, in the wall of a farm building near the church (1962); Cannock Chase, at the top of Abraham's Valley on the shale road going up towards the trig point (1969).

E. angustifolium L.   Rosebay Willowherb
Abundant in many habitats, but especially on waste ground and in woodland clearings: A–z, 778. 'Near the canal bridge at Oldbury, Staffordshire' (Withering 1796). During the nineteenth century this species was recorded for woods, a rabbit warren, ditch banks, gardens, cinder heaps and railway embankments in about thirty scattered places, but remained rare. In 1844 Garner recorded it for Whitmore and said that it spread so much in gardens that it was almost impossible to extirpate it. But in 1872 T. W. Daltry exhibited a plant to the North Staffordshire Field Club, which he had collected between Madeley and Whitmore railway stations, and said that he had not seen it anywhere else in that neighbourhood. This suggests that the plants Garner saw in the Whitmore gardens increased vegetatively rather than by seed. However, by 1901 Bagnall was able to say that it was frequent and 'apparently self-sown'.

Oenothera biennis L.   Common Evening-primrose
Waste ground: not systematically recorded. Occasionally as a garden weed (Brown 1863); Wolverhampton (Fraser 1882!); Birchills near Walsall (Bagnall 1901); semi-wild in gardens at Oakamoor (Berrisford MS); Shobnall (1956!).

Circaea lutetiana L.   Enchanter's-nightshade
Woods throughout the county: B, F–U, W–s, (t), u–w, y, (z), 228. Wood near Darlaston Hall (Forster 1796). (Map p 216).

C. intermedia Ehrh.   Upland Enchanter's-nightshade
Woods, mainly in the Churnet Valley: local. Dimminsdale (Carter 1839); woods at Froghall, Oakamoor and Heleigh Castle (Garner 1844); between Hanchurch and Clayton by a stream (Garner 1879);

Oakedge Park (Nowers 1910!); Highshutt near Cheadle (1942!);
Calwich Abbey (1947!); Hey Sprink, Madeley (1951!); Star Wood,
Oakamoor (1952!); Threap Wood (1955!); Marlpit Lane, Ellastone
(1963).

HALORAGACEAE

Myriophyllum verticillatum L.   Whorled Water-milfoil
Unconfirmed records for ditches on the north side of Aqualate
Mere (Dickenson 1798); railway pits south of Burton (Brown
1863); Kingston Pool (Fraser 1864); Perton Pool (Fraser 1864!).

M. spicatum L.   Spiked Water-milfoil
Recorded for canals, chiefly in the centre of the county, and less
often for ditches and reservoirs: (F), (N), P, S, W, (X), Y–Z, b,
(c–d), e–f, k–n, (r), (t), u–w, 29. Large drain on north side of
White-sitch Pool (Dickenson 1798).

M. alterniflorum DC.   Alternate Water-milfoil
Unconfirmed records for pool near Ingestre; Hopton Pools;
Sherbrook Valley; mill pool, Little Aston; pool at Oulton;
Foucher's Pool (all Bagnall 1901); canal at Lichfield (Druce in
BEC 1920 Rep).

HIPPURIDACEAE

Hippuris vulgaris L.   Marestail
Ponds in the south. 'About a mile from Stafford, on the foot-road
to Aston' (Withering 1787); Woodmill Brook (Gisborne 1791!);
road about half-way between Stafford and Newport (Clifford
1817); Kingston Pool, Rickerscote, Burton (Garner 1844); Perton
Pool (Fraser 1864!); pit at Willingsworth Furnaces (1950); Upper
Penn (1959); Short Heath, pond between canals on north side of
Wood Lane (1961!); pool east of Rushall Hall near Walsall (1962).

CALLITRICHACEAE

Water-starworts of one species or another are common throughout
Staffordshire in ponds, ditches and canals: B–u, w–y, 425. Pitt, 1796.
It is impossible to classify the old records satisfactorily without
specimens or until such specimens as exist have been critically
examined. But *C. stagnalis* is undoubtedly the commonest species.
The following account rests entirely on specimens in my herbarium
determined by P. M. Benoit in 1969, the word 'probably' indicating
a slight uncertainty.

Callitriche stagnalis Scop.   Common Water-starwort
Leekfrith (1948!); Aqualate Mere (1953!); Adbaston, margin of a lake (1954!); Chillington Pool (1954!); Heathylee, river near Hardings Booth (1955!); Fawfieldhead (1955!); Moss Carr (1966!).

C. platycarpa Kütz.
Forton (1954!), probably; Branston (1956!), probably; Croxden, watercress bed at Great Gate (1958!); Balterley Heath (1961!); Betley, in the main drain on Cracow Moss (1968!), very probably.

C. obtusangula Le Gall   Blunt-fruited Water-starwort
Drain near Yoxall Bridge (1956!), probably; Gnosall, Doley Common (1968!).

C. intermedia Hoffm.   Intermediate Water-starwort
River Dane, Cloud End House (1944!); Norton, feeder from Knypersley Pool (1949!); canal feeder below Rudyard Lake (1949!); pond at Cauldon (1958!); Wetton, fishpond at Back of Ecton (1968!); canal at Wall Grange (1969!).

C. hermaphroditica L.   Autumn Water-starwort
Lichfield, canal (1948!); Tixall, canal (1948!); canal at Tamhorn Park Farm (1968!); Freeford Pool (1968!).

LORANTHACEAE

Viscum album L.   Mistletoe
Bilston in the seventeenth century (Grigson 1958); apple trees near the Shropshire border (Pitt 1796); orchards near Four Ashes, Enville (Scott 1832); apple trees at Upper Arley (Watson 1837); Elmhurst and on a thorn bush at Rolleston (Garner 1844); Madeley and Great Haywood (Garner 1878); on a crab tree at Croxden (Berrisford MS undated); Rangemore, 'some apple trees full from top to near ground' (Nowers! undated); lane from Brereton to North Longdon (Reader 1923). Probably often introduced. 'I have met with only one instance of the mistletoe at all where it seemed to have grown independent of man, that was on the top of a tall poplar not far from Abbots Bromley' (Edwardes 1877).

CORNACEAE

Swida sanguinea (L.) Opiz   Dogwood
Woods and hedgerows except on acid soils: H–J, (K), L, N–P, S–j, m–s, (t), v–w, z, 131. Pitt, 1794. (Map p 216).

ARALIACEAE

Hedera helix L. Ivy
On trees, rocks, walls, in hedgerows and sometimes carpeting the ground in woods: A–z, 755. Plot, 1686.

UMBELLIFERAE

Hydrocotyle vulgaris L. Marsh Pennywort
Wet peaty places in fields, woodland rides and the beds of ponds from which the water has evaporated: frequent: B, E, G–H, K, (L), M–N, (P), R–U, W–a, (b), c–g, j–n, r–s, w, y, 69. Dickenson, 1798. (Map p 216).

Sanicula europaea L. Sanicle
In many of our richer woods: B, E, H–L, N–P, R–U, W–f, (g), j, n, r, t, v–w, z, 81. Near Yoxall Lodge (Gisborne 1792!). (Map p 217).

Chaerophyllum temulentum L. Rough Chervil
Common in hedgerows, especially in base-rich soils: B, E, H–w, z, 243. Near Yoxall Lodge (Gisborne 1791!). Map p 217).

Anthriscus caucalis Bieb. Bur Chervil
Recorded for waste places and 'hedges in a gravelly soil' (Dickenson 1798) in past days: (R), (Y–Z), (b), (f–g), (n), (r), (t), (w). The latest dated records are for roadsides near Baswich church (Reader 1920!) and near Colwich (Murray 1920!).

A. sylvestris (L.) Hoffm. Cow-parsley
Hedgerows, 'orchards and pastures' (Dickenson 1798): A–z, 728. Near Yoxall Lodge (Gisborne 1792!).

Scandix pecten-veneris L. Shepherd's-needle
A frequent cornfield weed in past days, but now rare: L, (N), P, (Y), (a–b), (g), (r), t, 3. Hadley End (Gisborne 1792!). West Bromwich, old marl pit (1948); Okeover, arable land by the Dove (1951); Newcastle, bird-seed alien (1970).

Myrrhis odorata (L.) Scop. Sweet Cicely
Common in the north on the banks of rivers and streams and in smaller but conspicuous colonies on grassy roadsides: (A), B–P, R, T–U, (Y), (b), 85. Tixall (Withering 1787). Known in Madeley churchyard for more than 100 years (1840! and 1951). (Map p 217).

Torilis japonica (Houtt.) DC. Upright Hedge-parsley
Hedgerows throughout the county: A, C–E, G–w, z, 330. Near Yoxall Lodge (Gisborne 1790!). (Map p 217).

T. arvensis (Huds.) Link   Spreading Hedge-parsley
Unconfirmed records for Betley, Stafford, High Offley, Uttoxeter
(Garner 1844); Burton (Brown 1863); Trysull (Fraser 1863!);
Perton (Fraser 1864!); Whittington, 1877 (Mathews 1884). It was
said to be not uncommon about Stafford and Burton (Garner 1882
and Brown 1863).

T. nodosa (L.) Gaertn.   Knotted Hedge-parsley
Rocks in Dovedale, Tutbury Castle hill, High Offley (Garner 1844);
near Wombourn (Fraser 1864!); Oakamoor (Berrisford 1902!);
Willoughbridge (Ridge in Flora); Burton (Curtis 1930).

Caucalis platycarpos L.   Small Bur-parsley
Casual. Burton (Nowers in Burton Flora 1901); Froghall
(Berrisford 1903!).

C. latifolia L.   Greater Bur-parsley
Casual. Burton (Nowers 1894!); Oakamoor (Berrisford 1902!);
railway embankment at Rocester (1907).

Coriandrum sativum L.   Coriander
Casual. Burton (Nowers 1894!); Oakamoor (Berrisford 1902!);
Froghall (Berrisford 1908!); North Street, Stoke (1931);
Springfields, Stoke (1940!).

Smyrnium olusatrum L.   Alexanders
'Occasional in and near old gardens, Endon' (Garner 1844).
Withering (1796) quotes Pennant: 'It was formerly cultivated in
our gardens, but its place is now better supplied by celery.' It has
recently (1971) been reported for a garden in Wolverhampton.

Conium maculatum L.   Hemlock
Canal sides and river banks: common between Wolverhampton
and Arley and abundant by the Dove and Trent: less common in
hedges near villages and on waste ground and rubbish tips: (A),
(G), J, L, (N), P, S–V, X–c, f–x, 74. Near the Dove (Pitt 1794).
(Map p 218).

Bupleurum rotundifolium L.   Thorow-wax
All the records are for waste ground and some of them should
probably be transferred to *B. lancifolium*. Near Newcastle High
School, Kitchener (1876); Stone (1894); Burton (Nowers 1902!);
Cheadle (1917); Alton (1919); Madeley (Daltry 1921!); Cannock
Chase (BEC 1922 Rep); North Street, Stoke (Ridge in Flora).

B. lancifolium Hornem. False Thorow-wax

Occurs frequently as a bird-seed alien, but is generally reported as *B. rotundifolium*. Oakamoor (Berrisford 1902!); roadside near Patshull (BEC 1917 Rep); Burton (BEC 1928 Rep); Yarnfield (1963); Newcastle (1970).

Apium graveolens L. Wild Celery

Salt brook, Shirleywich (Garner 1844 and Bagnall 1897!); Branston salt marsh (Brown 1863 and Nowers 1889!). Nowers and Wells (1890) found it abundant at Branston in two ditches. There are no recent records.

A. nodiflorum (L.) Lag. Fool's Water-cress

Common in ditches and shallow streams south of Stoke, but north of Stoke strangely rare except for Dovedale: E, J–y, 404. Dickenson, 1798. Riddelsdell and Baker (1906) record var *ochreatum* DC. for a marl pit at Huddlesford on the strength of a specimen collected by Power in 1832. There are no reliable records of *A. repens* (Jacq.) Lag., though creeping forms of *A. nodiflorum* were frequently so recorded.

A. inundatum (L.) Reichb. f. Lesser Marshwort

In shallow still water or in the mud at its edge: widespread, but uncommon: B, (F), (L), (W), X, (Y), Z, (c), (g), h, k–m, (p), t, (w), 7. North side of White Sitch pool (Dickenson 1798).

Petroselinum segetum (L.) Koch Corn Parsley

Unconfirmed records for cornfields at Tamworth (Garner 1844); Ecton Tor (Bloxam 1853); Warslow (Smith 1871); near Denstone (Edwardes 1876); marly banks, Hanbury (Bagnall 1901).

Sison amomum L. Stone Parsley

Hedgerows near Harlaston, as between Harlaston and Haunton (1946!) and in Mill Lane (1962!) and Portway Lane: very local. The only other record is a doubtful one for Castle Fields, Stafford (Moore 1897).

Cicuta virosa L. Cowbane

Ponds and fen ditches: local. 'The river Trent near Abbots Bromley' (Ray 1670); Kingston Pool (Withering 1787 and Bagnall 1897!); between Barton Mill and Borough End (Brown 1863); pond near Great Haywood (Murray 1920!); pond near Norbury (1953!); pond north of Waltonhurst near Eccleshall (1954!); Balterley Heath (1961!).

Ammi majus L. Bullwort

Madeley, in a field of mangolds near Bar Hill (Daltry 1916!); Burton (BEC 1930 Rep); Newcastle (1970).

Carum carvi L. Caraway

Unconfirmed records for Newcastle (Howitt in Watson 1837); Endon and roadside between Wolverhampton and Sedgley (Garner 1844).

Conopodium majus (Gouan) Loret Pignut

Woods, pastures, upland hayfields: A–z, 596. Wood near Rough Park (Gisborne 1791!).

Pimpinella saxifraga L. Lesser Burnet-saxifrage

Dry grassy banks and roadsides, especially on limestone and marl: (B), G–K, (L), M–P, R, T–U, X–b, d–h, (n), p–z, 95. 'Hills near Dudley Castle' (Wainwright in Shaw 1801). (Map p 218).

P. major (L.) Huds. Greater Burnet-saxifrage

Thickets, wood margins and shady hedgerows, especially on the Carboniferous limestone and Keuper marl, but with a curious distribution: locally plentiful in a triangle having Hollinsclough, Gnosall and Burton for its corners: C, G–J, (L), M–P, S–V, X–c, g, s, (t), (w), z, 131. Wednesbury (Withering 1796). (Map p 218).

Aegopodium podagraria L. Ground Elder

Damp shady places, hedgerows and neglected gardens, often forming extensive patches: A–z, 524. 'In some hedgerows and orchards in great abundance' (Pitt 1794).

Berula erecta (Huds.) Coville Lesser Water-parsnip

Pond margins, ditches and canal sides: frequent in the centre and south: J–K, R–S, W–Y, b, (c–e), f–t, w–x, 40. Kingston Pool (Stokes and Withering 1787). (Map p 218).

Oenanthe fistulosa L. Tubular Water-dropwort

Shallow ponds, ditches and water meadows: frequent in the south: (K), W–a, (b–c), d, (e–f), g–p, (r), s–t, 29. Near Yoxall Lodge (Gisborne 1792!). (Map p 219).

O. crocata L. Hemlock Water-dropwort

There are no recent records north of Lichfield, but south of Lichfield it is a common plant in ditches, on river banks and particularly along the canals: (N–P), (R), k–p, s–x, z, 36. Willoughbridge (Dickenson 1798). (Map p 219).

O. aquatica (L.) Poir.   Fine-leaved Water-dropwort
In shallow stagnant pools or in mud at the edge of the water:
frequent in central Staffordshire on the west side of the county
town, but rare elsewhere: E, L, (N), S, (T–U), (W), X–Z, (b), d, g,
j, (r), w, 19. 'Near Ranton Abbey by a pool' (Gisborne 1791!).

O. fluviatilis (Bab.) Colem.   River Water-dropwort
'Common in the Trent' (Brown 1863); Trent at Barton (Nowers
1892!). I have not seen the specimen.

Aethusa cynapium L.   Fool's-parsley
A common weed of gardens and cultivated fields: G, K–y, 216.
Near Yoxall Lodge (Gisborne 1791!).

Foeniculum vulgare Mill.   Fennel
Waste ground and railway embankments, particularly in the
Black Country: U, g, n, r–t, w–x, 10. Lichfield (1945!).

Silaum silaus (L.) Schinz and Thell.   Pepper-saxifrage
Water meadows and grassy roadsides: frequent about Stanton,
but rare elsewhere: (N), P, U, a, (b), d–e, (f), h, m, (r), t, y, 11.
Dickenson, 1798.

Angelica sylvestris L.   Wild Angelica
Marshy fields, stream sides and boggy woods: A–z, 618. Pitt, 1794.

Peucedanum ostruthium (L.) Koch   Masterwort
In hill pastures at Endon, Baddeley Edge and between Calton Moor
House and Mayfield (Garner 1844); naturalised at Folly (1947!);
riverside at Hardings Booth (1955). 'We have found it several times
in the ... moorlands ... but always near the site of some ancient
homestead ... and generally accompanied by the fragrant *Myrrhis
odorata*' (Garner 1855).

Pastinaca sativa L.   Wild Parsnip
Roadsides and waste ground: uncommon and often a relic of
cultivation, but long established on some of the limestone outcrops
in the south: (C), H, (K), (N), S, (T), (b), f, m, r–s, (t), w, 8. 'Near
Longnor in plenty' (Garner 1844).

Heracleum sphondylium L.   Hogweed
Abundant in upland hayfields and by roadsides: A–z, 795. Var
angustifolium Huds. frequently recorded. Stokes in Withering,
1787.

Daucus carota L.   Wild Carrot

Perhaps native in a few places, but most of the records are for waste ground, cultivated land and railway banks, and some of the plants recorded may be relics of cultivation: (F), H–J, (K), L, (N), R–T, (U), W–b, e–g, m–s, x, 38. Near Barton (Gisborne 1792!). 'Pastures in a marly soil, frequently so abundant as to be pernicious to the herbage in ill-cultivated lands' (Dickenson 1798).

CUCURBITACEAE

Bryonia dioica Jacq.   White Bryony

Common in hedgerows about Lichfield and south west of Wolverhampton: no recent records north of a line from Patshull to Tutbury: (K), (S), (Y), b, f–i, m–r, t–u, w–z, 122. 'Hedges near Lichfield' (Pitt 1794). (Map p 219).

EUPHORBIACEAE

Mercurialis perennis L.   Dog's Mercury

Woods and hedges, especially on limestone and marl: locally dominant: A–g, i–z, 504. Pitt, 1794.

M. annua L.   Annual Mercury

Garden weed at Burton (Nowers 1898!) and several later records for waste ground there.

Euphorbia lathyrus L.   Caper Spurge

Needwood (Hewgill in Garner 1844); weed in country gardens (Brown 1863); Cheadle (Masefield 1883!); Oaken (Fraser 1884!); Tatenhill, in old gardens (Nowers 1900!); garden weed, Oakamoor (Berrisford! undated); Clayton, garden in Northwood Lane (1949!); Stafford (1958); Stone (1958).

E. helioscopia L.   Sun Spurge

'In gardens and cornfields, common' (Pitt 1794): A–B, F–P, R–z, 304. This and the next species often grow together.

E. peplus L.   Petty Spurge

Gardens and arable fields: B, E–G, J–u, w–y, 300. Dickenson, 1798.

E. exigua L.   Dwarf Spurge

Cultivated land: less common than formerly: R, (T), Z, e–f, (g), r, t, 6. Pitt, 1794. Recent records for Mavesyn Ridware (1945!); Bobbington (1946!); cornfield near Loggerheads (1951!); West Bromwich, potato field north of Newton Road station (1954);

lane between Dunston and Penkridge (1957!); Abbots Bromley, arable field off Hobb Lane (1959!).

E. cyparissias L.   Cypress Spurge
Woods at Enville (Withering 1796); waste ground, Orgreave (Moore 1889!); Penkridge, abundant on the railway embankment east of Mansty Wood (1954!).

E. amygdaloides L.   Wood Spurge
Locally plentiful in Needwood Forest and Arley Woods, but rarely seen anywhere else: (N), T, (Z), a, v, z, 6. Near Yoxall Lodge (Gisborne 1792!). 'Very common on most parts of the Forest . . . I have seen this plant nowhere else wild in Staffordshire' (Pitt 1794). 'In the Holly Wood near Hardiwick' (Forster 1796), where it was rediscovered by T. J. Wallace in 1945. Bagnall (1901) records it for Star Wood, Oakamoor, but there are no other records for this wood. Seckley Wood (Fraser 1866! and Edees 1954!).

POLYGONACEAE

Polygonum aviculare L. sensu lato   Knotgrass
Arable and waste land: A–z, 777. Stokes in Withering, 1787. *P. aviculare* L. sensu stricto and *P. arenastrum* Bor. are both found but have not been separately recorded.

P. bistorta L.   Common Bistort
Wet meadows, often forming dense patches in a damp hollow: throughout the county, but commonest in the north: B–C, E–N, Q–U, X–a, (b), c–g, j–n, (p), r–t, v, y, 149. Essington (Pitt 1794). (Map p 219).

P. amphibium L.   Amphibious Bistort
Ponds, lakes, canals, rivers: B, D–y, 292. 'In Lush Pool' (Gisborne 1791!). (Map p 220).

P. persicaria L.   Redshanks
Arable land, sometimes abundant in fallow fields: A–y, 750. Dickenson, 1798.

P. lapathifolium L.   Pale Persicaria
Arable land: B–D, F–P, R–u, w–y, 357. Garner, 1844.

P. nodosum Pers.
Outwoods (Brown 1863); near Colwich (Bagnall 1901).

P. hydropiper L.   Common Water-pepper
'Rivulets and watery places' (Dickenson 1798) and on damp muddy
ground: A–w, y, 449.

P. mite Schrank   Tasteless Water-pepper
Unconfirmed record for Rudyard reservoir, W. T. B. Ridge (1934).

P. minus Huds.   Small Water-pepper
Wet muddy places by streams and ponds: rare. Wolstanton
(Garner 1844); Burton (Brown 1863); Branston (Nowers 1891!);
Slitting Mill near Rugeley (Reader 1923!); marshy field near
Wychnor church (1945!); Standon (1947); near Yoxall Bridge
(1956!); Barton (1956!); Doley Common near Gnosall (1958!);
Hopton Pools (1959); Barton, pond in a field at Holly Bank
Cottage (1962).

P. convolvulus L.   Black Bindweed
Cultivated land: A–D, F–z, 450. Pitt, 1794. Var *subalatum* Lej. and
Court. is recorded for Rushton (1944!).

P. cuspidatum Sieb. and Zucc.   Japanese Knotweed
Roadsides and waste places near gardens: now known throughout
the county: B, E–N, S–U, W–g, j–v, x–y, 143. Lichfield (Druce in
BEC 1923 Rep).

Fagopyrum esculentum Moench   Buckwheat
Rarely seen now, but once a common escape from cultivation.
'Mixed with barley on Heyley Castle Hill' (Plot 1686). 'In Staffordshire known only by the name of French Wheat. Much cultivated,
particularly in meagre soils. Cakes called crumpets are made of the
meal, which, toasted and buttered, are eaten with tea' (Dickenson
1798). 'Crops of buckwheat have been frequently raised on the
light lands near Kinver' (Scott 1832). About Cheadle (Carter 1839);
Castle Croft near Wolverhampton (Fraser 1883!); Hints (Bagnall
1901); Oakamoor (Berrisford 1902!); Church Eaton, a single plant
in a ride of High Onn Wood, perhaps from pheasant food (1952).

Rumex acetosella L.   Sheep's Sorrel
'Most abundant in a barren sandy soil' (Dickenson 1798): A–z, 637.

R. acetosa L.   Common Sorrel
Meadows, woodland glades and other grassy places: A–z, 775.
Pitt, 1794.

R. hydrolapathum Huds. Water Dock
Ponds, ditches and riversides, but most common along the canals:
E, G, K–M, R–U, W–b, d–u, w–x, 138. Tamworth (Withering
1787). (Map p 220).

R. alpinus L. Monk's Rhubarb
By roadsides near farms and old buildings in many places in
the moorland area: B–C, G–H, K, M–N, R, T, (U), Y, e, (j), 36.
Roches district (1868). (Map p 220).

R. crispus L. Curled Dock
Arable and waste land: A–z, 639. Pitt, 1794. *R. crispus* x *obtusifolius*
has been recorded occasionally and is probably common.

R. obtusifolius L. Broad-leaved Dock
Roadsides, field borders and disturbed ground: B–z, 782. 'Rubbish
and pastures; deer eat it with avidity' (Dickenson 1798).

R. pulcher L. Fiddle Dock
Unconfirmed records for Cheadle (Bourne in Dickenson 1798);
Stoke (Garner 1844); Burton (Flora 1901).

R. sanguineus L. Wood Dock
Woods and hedgerows: A–C, F–x, z, 331. Near Cheadle (Bourne
in Dickenson 1798).

R. conglomeratus Murr. Clustered Dock
Ditches and wet places by the sides of rivers, canals and ponds:
B, G, J–z, 283. Tamworth (Withering 1787).

R. maritimus L. Golden Dock
In mud at the edges of ponds and lakes in the south of the county:
rare: (Y), Z, (b), d, (i), j, (m), p, r, w, 9. Dam of Kingston Pool
(Garner 1844). The recent records, some of them confirming
nineteenth century records, are as follows: Foucher's Pool (1946!);
pond half way between Church Eaton and Gnosall (1952!);
Belvide reservoir (1954!); Chillington Pool (1954!); Pool Hall
Pool (1954!); canal side at Brewood on dredged mud (1957!);
pond south of The Hollies, Enville (1960!); Warwickshire Moor,
Tamworth (1962); Blithfield reservoir (1964).

URTICACEAE

Parietaria judaica L. Pellitory-of-the-wall
'On old walls, on the gateway leading to Lichfield minster' (Pitt
1794): (L), P, (U), W, (Y–Z), b–c, g, j, n–r, v–w, 18.

Urtica urens L.   Annual Nettle
Arable fields, manure heaps, farmyards, tips and waste ground:
C, G–u, w, y, 180. Dickenson, 1798.

U. dioica L.   Common Nettle
Waste ground, hedgerows and wet places in woods: A–z, 792.
Pitt, 1796.

CANNABIACEAE

Humulus lupulus L.   Hop
Hedges, especially near villages: B, D–G, K–U, W–w, y–z, 156.
Whittington (Pitt 1794). Garner (1844) said it was once cultivated
in the south west of the county. (Map p 220).

Cannabis sativa L.   Hemp
Casual. Cornfield at Alton (Berrisford 1910!); Patshull (BEC 1923
Rep); Lichfield (Harlond 1938!); Burton (Burges 1944).

ULMACEAE

Ulmus glabra Huds.   Wych Elm
Woods and hedgerows throughout the county, but particularly
common in the Manifold Valley: B–z, 634. Pitt, 1796. There was a
famous elm at Tutbury known in the early seventeenth century as
the Dun's Cross Elm and later as the Big Elm. In 1818 its branches
were said to spread 108 feet and its trunk to be nearly seventeen
feet in circumference at six feet from the ground. It was sixty feet
tall. A drawing by Garner shows it to be a decaying tree in 1844
and by 1849 only the stump was left (Garner MS). Brown (1863)
tells us that it was destroyed by fire from a gipsy encampment.
From old maps we can define its position as SK/212281.
    The following hybrids have been determined by R. Melville:
U. glabra x *carpinifolia*: Whitgreave (1944!); Keele, in the churchyard against the church on the north side (1958!); Seighford,
roadside north of the bridge at Butter Bank (1958!): *U. glabra* x
plotii: Whitgreave (1944!); Barrowhill, Rocester (1958!); Blithfield,
lane from Admaston to Steenwood Cottages (1956!): *U. glabra* x
carpinifolia x *plotii:* Tixall, Brancote Covert on the western side
(1956!); canal from Kennels Farm (1956!); Hanyards Lane (1956!);
Hardiwick Heath, Sandon (1956!): *U.* x *hollandica* Mill.: Butterton
near Whitmore (1944!).

U. procera Salisb. English Elm

A common hedgerow tree in the south of the county: (B), K–L, U, W–Z, b–x, 140. Himley (Withering 1787). (Map p 221).

U. plotii Druce Small-leaved Elm

Near Lichfield (Druce in BEC 1919 Rep); near Stafford (BEC 1923 Rep); Keele, SJ/806447 (1958!), det Melville; Swynnerton, about 100 yards on the Newcastle side of Clifford's Wood Farm (1958!), det Melville. The following were also determined by Melville: *U. plotii* x *carpinifolia*: Little Bridgeford, SJ/875276 (1958!); Broad Heath (1958!); Ranton, between Five Lanes Ends and Stubbs Wood (1958!): *U. plotii* x *coritana*: Broad Heath, SJ/854254 (1958!): *U. plotii* x *coritana* x ? *glabra*: Moreton Grange, Colwich (1956!).

MYRICACEAE

Myrica gale L. Bog Myrtle

Today found only on Loynton Moss near Norbury, where it survives in diminished quantity. In 1901 Bagnall said it was abundant on Norbury Big Moss, which was part of Loynton Moss, and the moorlands near Norbury. There is a specimen in his herbarium gathered in 1894. In 1969 it was still plentiful. The other records are as follows: 'A very aromatic shrub of the myrtle kind . . . grows spontaneously at a place called Foume, or Foulmere, near Cannock Wood, about a mile from the Four Crosses. It is called gale, or sweet gale, and gives name to a small hamlet near it. Where it flourishes is a big morassy ground, between two coppices, greatly sheltered from the bleak winds. . . . It seems confined to this small spot of a few acres' (Observator 1786 for 1784). 'On the north side of Aqualate Mere' (Withering 1787). 'Forton Moors, near Aqualate Mere, Moreton Moors, in great abundance' (Dickenson 1798). A bush discovered on the Downs Banks, Stone, in 1947 was later found to have been planted.

BETULACEAE

Betula pendula Roth Silver Birch

Throughout the county, especially in dry gravelly soils: A–z, 545. Plot, 1686.

B. pubescens Ehrh. Downy Birch

Often in wetter places than *B. pendula* and the commoner species of the moorlands: B–C, E–H, K–U, W–X, Z–g, j–r, t–u, w–y, 180. Brown, 1863.

Alnus glutinosa (L.) Gaertn. Alder
Streamsides and marshy woods: A–z, 695. Stokes (Withering 1787) tells us that it was called owler in Staffordshire.

CORYLACEAE

Carpinus betulus L. Hornbeam
Hedgerows and woods: frequent, but not native: C, G, L, (N), Q–U, W–d, f–g, j–k, n, r, t, w, y, 41. Pitt, 1796.

Corylus avellana L. Hazel
Copses, hedges, limestone hillsides: A–z, 683. Pitt, 1796.

FAGACEAE

Fagus sylvatica L. Beech
'Common in hedges and plantations' (Pitt 1796): B–z, 453.

Castanea sativa Mill. Sweet Chestnut
Parks and plantations: frequent: B–C, E–U, W–g, i–r, t–w, z, 145. Pitt, 1796. 'The nuts . . . are roasted in small stoves in the streets by the fruit women and in winter form a very agreeable addition to our deserts' (Withering 1787).

Quercus robur L. Pedunculate Oak
Woods and hedgerows: A–C, E–z, 705. 'Needwood Forest in considerable quantity' (Pitt 1794). Dickenson tells us that in 1798, 'notwithstanding the pressing demands of modern times', about 1,000 acres of oak still enriched the Forest and that in the clayey and marly lands the hedgerows were full of self-sown oaks. In our ultra modern days oak woods all over the county have given place to fir plantations.

Q. petraea (Mattuschka) Liebl. Sessile Oak
This species prefers the drier acid soils of the Bunter sandstone and Millstone grit: insufficiently recorded: B, (F), H, N, R, (S), U, Y–Z, (a), (f), k, (n), (r), t–u, (w), y–z, 21. 'Very common about Himley and Kingswinford' (Purton 1821).

SALICACEAE

Populus alba L. White Poplar
Frequent as a planted tree: (F), G, J–K, (L), M, (N), P, R–U, W–b, d, (e), g, j–n, r, t–y, 51. Pitt, 1796.

P. canescens (Ait.) Sm. Grey Poplar

Probably frequent, but records have not been kept. 'Large trees at Handford Bridge' (Garner 1844).

P. tremula L. Aspen

Damp woods and hedges: frequent: C, F–H, K–P, R–U, W–a, c–g, j–r, t–x, 93. 'Succeeds best in moist situations' (Pitt 1796).

P. nigra L. Black Poplar

The true species with drooping branches, large bosses on the trunk and attenuate leaves occurs here and there by streams and roadsides in the south and west: Q, S, X, c, e–f, j–k, r, v–w, 13. Early records uncertain. Gnosall, several trees by Doley Brook south of the railway, SJ/832202 (1958!); Upper Arley, roadside above Brittle's Farm (1959!); Kinver, on the county boundary near Iverley House Farm (1959); Wrottesley, Perton Lane (1961!); Codsall, in the field near Bilbrook, SJ/876040 (1961!).

P. x canadensis Moench Italian Poplar

This is the commonly planted tree of woods and hedgerows which is usually called Black Poplar: B, E–H, K–N, R–U, W–y, 150. Brown, 1863.

Salix pentandra L. Bay Willow

Native in the north of the county where it is common by moorland streams: B–D, F–K, (L), M–P, R–U, W, Y, (f), j, (m), (r–s), t, w, y, 71. Wolverhampton, near a stable, 1639 (Johnson 1641). This is both the first record for Staffordshire and also for Britain. (Map p 221).

S. alba L. White Willow

By rivers and streams: common in the south, rare in the north: L, N, S–V, Z–b, d–g, i–m, p–r, (t), (v), w–y, 45. Pitt, 1796. Many of the Staffordshire trees have fragile twigs and leaves which are only sparsely hairy on the upper surface at maturity. One of these growing near Yoxall Bridge (1956!) was determined as var *caerulea* (Sm.) Sm. by R. D. Meikle, though he had only a spray of leaves to judge from. A small tree with very silky leaves found in Sandon Wood (1956!) may be var *regalis* Hort. Var *vitellina* (L.) Stokes was recorded for Gailey reservoir by Druce (BEC 1923 Rep).

S. fragilis L. Crack Willow

'Hedges and banks of rivers' (Dickenson 1798): A–y, 608.

Dickenson continues: 'A very large tree of this species, growing in a field adjacent to the city of Lichfield, bears the name of Dr Johnson's willow. It was not planted by the venerable sage, but he delighted to recline under its shade.' Garner (1844) gives us a long account of this tree and tells us that it perished in a violent storm on 29 April 1829 at 3 pm.

S. triandra L.   Almond Willow
Riversides and old osier beds, often of planted origin: rare. Osier beds below Stoke (Garner 1844); Burton (Brown 1863); Pendeford (Fraser 1864); Branston (Burton Flora 1901); near Colton (Bagnall 1901); Hollybank Lane near Brereton (Reader 1922!); marshy field near the church at Wychnor, male bushes in full flower on 30 August (1945!); Uttoxeter goods yard, abundant, all female bushes (1953); Windswell Pool, Forton (1964!). A specimen collected at Bilbrook in 1881 was determined as *S.* x *trevirani* Sprengl. by J. Fraser (BEC 1926 Rep).

S. purpurea L.   Purple Willow
Osier beds, ditches and river banks: frequent: B, E, K, (L), M–N, S, V–W, Y, a–b, g, (i–j), (m), n, (r), w, 20. Hanford (Garner 1844). The common variety is var *lambertiana* (Sm.) Koch, but Garner (1844) records var *helix* (L.) Koch from the Trent at Stoke, where it used to be frequent, and J. Fraser so determined a specimen from Bilbrook collected by an earlier J. Fraser in 1888 (BEC 1926 Rep).

S. daphnoides Vill.
One unconfirmed record for Gailey (Druce in BEC 1923 Rep), 'a pleasing feature round the reservoir' (BEC 1924 Rep, p 453).

S. viminalis L.   Osier
By the sides of rivers, ponds, streams and in marshy woods: B–D, G–N, R–W, Y–z, 111. Pitt, 1796. Var *linearifolia* Wimm. and Grab. is found occasionally.

S. calodendron Wimm.
Trysull, Wolverhampton, W. and J. Fraser, 1873! and 1878! (Meikle 1952); Cannock Chase, 1961 (R. C. L. Howitt in litt); Marchington, roadside between Brook House Farm and Brookside Farm (1964!), det Meikle.

S. caprea L.   Goat Willow
Copses and hedges: B–C, E–n, r–z, 288. 'No species of *Salix*

requires such a dry soil as this does' (Withering 1796). 'Common in hedges, may be known by its large roundish leaves white below' (Garner 1844).

S. x *laurina* Sm. (*S. caprea* x *viminalis*)
Gnosall, SJ/8319 (1961!); Uttoxeter, near Newlands (1964!); Marchington, Moisty Lane (1964!); all three det Meikle.

S. cinerea L. subsp *oleifolia* Macreight   Grey Willow
'This is perhaps the most common of all our willows, as well in hedgerows as in woods, . . . but not in a dry soil' (Withering 1796): A–z, 495. Dickenson, 1798.

S. x *forbyana* Sm. (*S. cinerea* x *purpurea* x *viminalis*)
Staffordshire bank of the Dove, just below Lode Mill, several old trees about fifteen feet high (Purchas 1885); Rolleston, south bank of the Dove east of Dove Bridge (1964!), det Meikle.

S. x *smithiana* Willd. (*S. cinerea* x *viminalis*)
Hedges and streamsides: frequent: C, F, H, K, R–S, V–Y, d, f–g, k, w, 23. Near Bilbrook (Fraser 1881!).

S. aurita L.   Eared Willow
Moors west of Flash (1964!), det Meikle, and Swallow Moss; Staffordshire bank of the Dane, between Ludchurch and the old mill (Howitt 1965!): probably frequent on the moors. Blymhill (Dickenson in Withering 1801), but no specimen to support the record. *S. aurita* x *cinerea*: Leigh, roadside between Bearsbrook and Godstone (1964!), det Meikle.

S. repens L.   Creeping Willow
Wet places on heaths and moors: local. 'South corner of a bog about 200 yards south west of the Hawk Hills bridge over Woodmill Brook, 1792. In the gorse half-way between the corner of the Foxholes and Whitemere Bog' (Gisborne MS); Pendeford (Pitt 1796); Aqualate (Dickenson 1798); Goldsitch Moss (Bloxam 1853); Cheadle (Masefield 1883!); Sherbrook Valley (Bagnall 1901); near Stanton (Ridge in Flora); Brocton (1948!), det Meikle; Moss Carr (1949!); Ipstones Edge (1953); southern end of Deep Hayes reservoir (1955!); Gun near Leek (1969).

ERICACEAE

Rhododendron ponticun L.   Rhododendron
Planted in woods and shrubberies and regenerating freely: B–C, F–U, W–g, i–r, t–x, 193. No published records.

Andromeda polifolia L.   Bog Rosemary
Upland peat bogs and lowland mosses: rare or very local. Chartley
Moss (Bagot in Withering 1801); Whitmore Heath and Craddocks
Moss (Garner 1844); Goldsitch Moss (Bloxam 1853); Warslow
(Smith 1871). It survives on Chartley Moss and in small quantity
on Craddocks Moss and Goldsitch Moss.

Calluna vulgaris (L.) Hull   Heather
Heaths and moors: A–D, F–U, W–X, Z–a, e–g, k–n, r, t–x, z, 217.
Plot (1686) tells us that the plains or hays of Cannock Chase below
the wooded summits of the hills were in great part covered only
with the purple odoriferous ling. (Map p 221).

Erica tetralix L.   Cross-leaved Heath
Common in moorland bogs and in wet places on Cannock Chase:
B–C, E, (F), G–P, R–U, W, Z, e–f, k–m, (n), r, t, w, 60. Stokes in
Withering, 1787.

E. cinerea L.   Bell Heather
Dry places on heaths and moors, particularly in the Churnet
Valley, on Cannock Chase and about Kinver: B–C, (F), G–H,
K–N, R–U, (Y), e–f, m, (n), r–t, w, z, 45. Stokes in Withering, 1787

Vaccinium vitis-idaea L.   Cowberry
Acid heaths and moors, often growing with *V. myrtillus* but less
common: most plentiful on Cannock Chase: B–C, (F–G), H, K,
(L–M), N, S–T, Z, e–f, (m), 29. 'Upon boggy mountains and heaths
in Staffordshire' (Ray 1670). 'On the summit only of the most
pointed of the Camp hills, in a dry gravel' (Waring 1770). Still there
in 1970.

V. myrtillus L.   Bilberry
Heaths and oak-birch woods on the Bunter sandstone and Millstone
grit: B–D, F–P, R–U, Y–a, e–f, m–n, t, w–x, z, 193. 'One part of it
(Wolseley Park) is pretty full of bilberries which thrive under the
shade of the oaks' (Celia Fiennes 1698). It was already a long
established custom for the country people to make 'booths and a
sort of fair' outside Wolseley Park in the bilberry season and
gather the fruit for sale. Wright (1934) tells us that in the middle of
the nineteenth century bilberries gathered on Cannock Chase were
sold in Stafford market at 3d a pint. (Map p 221).

V. myrtillus x *vitis-idaea*   Hybrid Bilberry
The hybrid bilberry was first discovered in Britain on Maer Hills

(Camp Hills) by D. Ball in 1870. It was growing with *V. myrtillus* and *V. vitis-idaea* and seemed to be intermediate between them. Specimens were submitted to Garner, who sent one to Charles Darwin and exhibited another to the North Staffordshire Field Club at a meeting held in Newcastle on 23 February 1871. Garner told the members that he had received a reply from Darwin that very afternoon expressing pleasure that the specimen came from a place to which he was so much attached and suggesting that attention should be paid to the fruit, because, if the plant were a true hybrid, the fruit would be nearly or quite sterile. On 7 March 1872 Garner exhibited specimens to the Linnean Society and the following year published an illustrated description in *Science Gossip*.

In August 1886 T. G. Bonney found the same plant on Cannock Chase growing plentifully with *V. myrtillus* and *V. vitis-idaea* and in 1887 N. E. Brown, not knowing that it had already been found at Maer and that Garner had published a description of it, issued a fresh account with the title, '*Vaccinium intermedium* Ruthe, a new British Plant.' This too was accompanied by an excellent illustration. Thereupon Garner wrote to Brown and informed him of the earlier record. Brown saw the specimens Garner had exhibited to the Linnean Society and agreed that they were identical with those from Cannock Chase.

There are many subsequent records for Cannock Chase. In 1919 W. B. Gourlay and G. M. Vevers found it in a score of distinct and widely separated localities on the Chase, especially where the heath had been broken up by the soldiers training there during the Great War. It was locally very abundant and one patch, centred on an artificial bank crowned with birch trees, was about half an acre in extent. In 1950 V. Jacobs found a dense patch in Brindley Valley north of Hednesford and it can be seen in several places near Broadhurst Green. Fruit is sparingly produced.

There are other records as follows: Oulton, canal embankment (Bagnall 1896!); Norton Bog (Bagnall 1898!); Whitmore Common (Gourlay 1919); near the Bridestones (Daltry 1946!); between Oakamoor and Whiston Eaves (1947); Ipstones Edge in 1961; Hanchurch Hills, SJ/837401 (1966).

V. oxycoccos L.   Cranberry
Lowland mosses and peat bogs among the hills: local: B–C, E,

(F–G), K, M–N, R–S, (W), Z, (c), e–f, m, 20. 'The Marrish Whortle groweth upon bogs and such like waterish and fennie places, especially in Cheshire and Staffordshire, where I have found it in great plentie' (Gerard 1597). There are old records for Bishop's Wood, Cannock Chase, Aqualate Mere, Weston-under-Lizard, Chartley Moss, Norton Bog, Norbury Big Moss and many places in the north. Today it is most plentiful on Chartley Moss, where it flowers and fruits freely.

PYROLACEAE

Pyrola minor L.   Common Wintergreen
Cotton Dell and Dimminsdale, by the side of the private coach road (Carter 1839), said to be commoner in Dimminsdale than in Cotton Dell (1877), specimens from Dimminsdale collected by Masefield (1883!) and Berrisford (1903!), last seen by Pickard (BEC 1925 Rep); Ramshorn (Ridge 1915!).

P. rotundifolia L.   Round-leaved Wintergreen
Wood near Cotton Hall (Dickenson 1798), probably a mistake for *P. minor*. Old records for Chartley Moss from 1801 (Bagot in Withering) to 1886 (Tylecote), but no specimens.

MONOTROPACEAE

Monotropa hypopitys L.   Yellow Birdsnest
Unconfirmed records for woods at Enville (Withering 1787) and Gospel End (Wainwright in Shaw 1801).

EMPETRACEAE

Empetrum nigrum L.   Crowberry
'On the moors and fells in Staffordshire' (Ray 1670), 'both in the driest and most barren rocky soils and in bogs and moorish grounds' (Withering 1787): B–C, E, (F), G–H, (K), (M), N, (R), S, (T), Z, e–f, (m), 25. 'Abundant on many parts of Cannock Chase; its glowing purple berries are a beautiful ornament to those dreary tracts' (Dickenson 1798). Occurs also on Hanchurch Hills, Balterley Heath, Chartley Moss and in Ousal Dale.

PRIMULACEAE

Primula veris L.   Cowslip
Rough bushy hillsides, especially in Needwood Forest and on the limestone, railway embankments and more rarely in meadows:

(*above*) The Weaver Hills, home of the Mountain Pansy (*Viola lutea*) and the Grass-of-Parnassus (*Parnassia palustris*); (*below*) Hawksmoor Wood, where *Rubus bellardii* grows

(*above*) Hedge Bindweed (*Calystegia sepium*); (*below*) Large Bindweed
(*Calystegia silvatica*)

B, E, (F), H–P, R–U, W–X, (Y), Z–a, c–d, f–j, m, p, (r), s, u–x, 100. Pitt, 1794. *P. veris* x *vulgaris*, the false oxlip, is frequently seen where cowslips and primroses grow together. (Map p 222).

P. elatior (L.) Hill   Oxlip
Naturalised in the churchyard at Alstonfield (1960!).

P. vulgaris Huds.   Primrose
Damp hedge bottoms and by streams in woods on limestone and marl: B–D, (F–G), H–L, N–a, c–k, (r), t, v–w, z, 221. Pitt, 1794. (Map p 222).

Hottonia palustris L.   Water-violet
Fen drains and ponds: very local: E, (L), (W), (Y), (g), h, (p), (t), 2. 'In full flower abundantly, 3rd May 1794, by the roadside from Barton Turn to Newbold and from Branston to Burton' (Gisborne MS). In 1956 it was seen in one of the lily pits at Branston, SK/221197. 'In a ditch in very boggy ground on the south side of Aqualate Mere' (Dickenson 1798). 'In a muddy ditch near the navigation bridge at Fazeley' (Power MS). Balterley Heath (1961!).

Lysimachia nemorum L.   Yellow Pimpernel
Moist places in woods: A–C, F–U, W–f, j–k, (m), n, r, t–w, z, 159. Belmont (Sneyd in Pitt 1794). (Map p 222).

L. nummularia L.   Creeping Jenny
'Marshy and boggy meadows, frequent' (Dickenson 1798): B, (G), J, (L), M–N, S–b, d–p, s–v, y, 105. (Map p 222).

L. vulgaris L.   Yellow Loosestrife
Watery lanes, canal sides and marshy woods: rare or very local as a native plant, most of our records being for waste ground near gardens: B, G–H, (L), (N), (R–S), W, (X–Y), d–e, n, 9. 'Blymhill, in Mutty meadow hedge' (Dickenson 1798). Grows with *Ranunculus lingua* on Loynton Moss in a natural habitat.

L. punctata L.   Dotted Loosestrife
Introduced. Checkley, between the ponds on the west side of Greatgate Wood, SK/046407 (1958!); Haughton, damp lane to Woodhouse Farm, SJ/854213 (1958!).

L. thyrsiflora L.   Tufted Loosestrife
Newborough, swampy ground at the southern end of the lake in Holly Bush Park, Miss D. Meynell (1952!).

Anagallis tenella (L.) L. Bog Pimpernel
Wet peaty fields and bogs on lowland heaths: chiefly on Cannock
Chase and near Stone and Eccleshall: R–S, (W), (Z), (c), e–f, m, 10.
'Tittensor Hills and Cannock Heath' (Stokes in Withering 1787).

A. arvensis L. Scarlet Pimpernel
Arable fields, especially in light soils: common in the south:
H, K–N, R–T, W–y, 189. Pitt, 1794. (Map p 223).

A. minima (L.) E. H. L. Krause Chaffweed
Unconfirmed records for Blithfield (Bagot in Withering 1801) and
Balterley (Writtle in Garner 1878).

Glaux maritima L. Sea-milkwort
Salt marsh near Tixall (Wolseley in Dickenson 1798) and at
Ingestre (Bagot in Withering 1801). Clifford (1817) was afraid that
the immense drains made by Lord Talbot in the salt marsh at
Ingestre would destroy the marine plants, but some of them
survived. Meadow between Shirleywich and Ingestre (Reader
1923!); Pasturefields, SJ/992248 (1956!).

Samolus valerandi L. Brookweed
Wet muddy ground in central Staffordshire: rare. Tixall salt marsh
(Clifford 1817); Mott's Well near Smallwood Manor (Garner
1844); Aqualate Mere, south side (Fraser 1863); Branston salt
marsh (Nowers 1889!); remains of salt marsh near Tixall (Reader
1923!); salt marsh near Ingestre (Thornton 1925!); with *Anagallis
tenella* in a rushy meadow near Stone, SJ/881332 (1948!); plentifully
near the boathouse on the south bank of Aqualate Mere (1953!);
one plant in a drain on Allimore Green Common (1968).

OLEACEAE

Fraxinus excelsior L. Ash
Woods and hedgerows throughout the county, but most abundant
on the limestone: A–z, 768. Plot, 1686.

Ligustrum vulgare L. Wild Privet
Hedges and thickets: native in base-rich soils, but often planted:
G–L, (N), P, R–U, W–s, w–x, z, 96. 'Between Ivetsea-bank and
Cannock Heath' (Waring 1770).

APOCYNACEAE

Vinca minor L. Lesser Periwinkle
Hedgebanks and thickets: (E–F), K, (L), M, (N), P, R, T–U,
W–X, (Y), Z, (b), (h), j, m, r, (t), w–x, (y), 19. Pitt, 1794.

V. major L.   Greater Periwinkle
Several old records for hedgebanks near houses. Pitt, 1794.

GENTIANACEAE

Centaurium erythraea Rafn   Common Centaury
Dry banks, wood clearings, gravel pits and open places on sandstone heaths: frequent: (F), H, K–M, (N–P), Q–U, W–Z, (a–b),
d–f, (g), j–m, r–s, v–w, y, 49. Pitt, 1794. (Map p 223).

Blackstonia perfoliata (L.) Huds.   Yellow-wort
Dry banks and quarries on the Silurian limestone and as a rare
casual along the canals: (K–L), (N), (R), W, (X), d, s, (t), (w), 4.
Ranton Abbey (Withering 1787). 'In great abundance upon the
lime hills near Dudley Castle' (Wainwright in Shaw 1801).

Gentianella campestris (L.) Börner   Field Gentian
Recorded for dry pastures in the north, but the early botanists may
have mistaken forms of *G. amarella* for *G. campestris*: (B) C,
(G–H), (N), (Z), 1. 'On bur walls, near Wolseley Bridge' (Wolseley
in Dickenson 1798). The only recent record is for Coatestown,
Hollinsclough, R. H. Hall (1949).

G. amarella (L.) Börner   Autumn Gentian
Limestone pastures: local in the north, as on Ipstones Edge and in
the old quarries at Caldonlow: C, H, N–P, (s), 7. Bourne in
Dickenson, 1798.

MENYANTHACEAE

Menyanthes trifoliata L.   Bogbean
Pits, reed swamps and wet moorland bogs, often with *Potentilla
palustris:* local: (B), C, F–H, (K), M–N, (P), R, T–U, W, (X),
Y–Z, (a–b), (d), f, (g), m–n, (w), 23. 'In the pit in the Foxholes and
in Whitemere bog, 1792' (Gisborne MS). (Map p 223).

Nymphoides peltata (S.G.Gmel.) Kuntze   Fringed Water-lily
'Sometimes introduced into ponds, but apparently not indigenous
in the district' (Brown 1863). Pitt (1794) said it was common in
pits, but he was surely mistaken. Introduced into a pond at Alton
(Garner 1844). In the canal (now drained) between Forton and
Norbury (1944).

POLEMONIACEAE

Polemonium caeruleum L.   Jacob's-ladder
'Plentiful on the north aspect of limestone precipices' (Garner

1844): frequently recorded as a garden escape, but as a native plant confined to the Carboniferous limestone. Wetton (Carter 1839); Thor's Cave and Ecton Hill (Garner 1844); Peasland Rocks in Wolfscote Dale, many scores of plants (1953!). Continues to thrive on rock ledges about Thor's Cave.

BORAGINACEAE

Cynoglossum officinale L.   Hound's-tongue
Dry sandy or calcareous soils: locally common in former days, but now rare. Pitt, 1794. Lane between Tixall and Hoo Mill (Clifford 1817); frequent on coalpit mounds, Betley, Madeley, Burton, Dovedale (Garner 1844); Outwood Hills, Tatenhill (Brown 1863); Wren's Nest, Aqualate (Fraser 1864); Bunster Hill (Smith 1871); near Shelton lime kilns (Garner 1879); waste ground near Armitage (Reader 1919!); Loynton Moss (1933 and 1948!); Forton, about rabbit holes on a sandy knoll at the west end of Aqualate Mere (1954!); one plant at the foot of a small quarried cliff by the Branston road at Tatenhill in 1966 (1968).

Asperugo procumbens L.   Madwort
Burton, on malthouse rubbish (Nowers 1902!); Oakamoor (Berrisford 1902!); Rocester, railway embankment (1908); North Street, Stoke (Ridge in Flora).

Symphytum officinale L.   Common Comfrey
Native in a few places in the south of the county in swampy woods and by streams, but usually found as an established alien by roadsides and on waste ground near houses: B–C, E–H, K–P, R–V, X–Z, 122. These records include plants with purple-tinged flowers, which may be back-crosses with *S. x uplabdicum*, but all had strongly decurrent leaves. 'Ditches and banks of rivers' (Dickenson 1798). Specimens in my herbarium, det A. E. Wade, from Balterley, Brewood, Keele, Lapley, Lichfield, Longdon, Northwood, Ranton and Warslow.

S. x *uplandicum* Nyman
Commonly found outside cottage gardens and in village hedgerows: B–C, G–H, K–T, W–Y, a–b, e–g, j–x, z, 94. Aldridge (Druce in BEC 1926 Rep), but records of *S. asperrimum* for near Ashbourne (Purchas 1879) and near Brereton (Reader 1920!) probably belong here. Specimens in my herbarium, det Wade, from Anslow,

Brewood, Codsall, Forton, Kinver, Longsdon, Newborough and Whitmore.

S. tuberosum L. Tuberous Comfrey
Unconfirmed and uncertain records for Lichfield (Jackson 1837); Wetton Valley, Longnor (Garner 1844). One recent record for Longdon, near the stream in Beaudesert Old Park, SK/0514 (1952!), conf Wade.

Borago officinalis L. Borage
Garden escape. Cheadle (Bourne in Dickenson 1798); Lichfield (Jackson 1837); Needwood, Burton, Dudley Castle (Garner 1844); Codsall (Fraser 1876!); Burton (Nowers 1905!).

Pentaglottis sempervirens (L.) Tausch Green Alkanet
Naturalised in a few places on hedgebanks near houses: (N), S, W–X, (Y), a, e–f, r, (t), (x), y, 10. 'Among the old ruins at Tixall' (Clifford 1817). Known at Acton near Stafford for many years (Moore 1897 and, roadside near Acton Hill, 1954!).

Lycopsis arvensis L. Bugloss
Sandy arable fields, especially near Eccleshall and Lichfield and in the south west: K–M, (N), R–S, (T–U), W, (Y), b–c, e–g, j–n, r–t, w, v, 68. Pitt, 1796. (Map p 223).

Myosotis scorpioides L. Water Forget-me-not
Wet places by rivers, streams and canals: A–D, G–b, d–r, t–y, 244. Forster, 1796.

M. secunda A. Murr. Creeping Forget-me-not
By rills and streams in acid peat: plentiful on the moors and on Cannock Chase, but uncommon elsewhere: B–C, G–H, M–N, R–S, e–f, 34. Garner, 1844. (Map p 224).

M. caespitosa K. F. Schultz Tufted Forget-me-not
Marshy fields and wet muddy places by ponds and streams: B–t, w, 350. Lichfield (Jackson 1837).

M. sylvatica Hoffm. Wood Forget-me-not
Moist loamy woods, especially on the Carboniferous limestone: C–D, H–P, R–U, W, Y, a–d, f, j, t, v–w, 74. Yoxall Lodge, 1837 (Babington 1897). Garner (MS) tells us that he saw it in flower at Heleigh Castle on 7 April 1852: I saw it there in full flower on 19 May 1954. (Map p 224).

M. arvensis (L.) Hill   Field Forget-me-not
Arable fields, railway banks, wall tops, woods: A–B, D, G–x, z,
328. Pitt, 1794.

M. discolor Pers.   Changing Forget-me-not
'In dry sandy places, and in somewhat boggy meadows' (Purton
1821): (B), (F), K–L, N–P, R–T, W–Z, (a–b), d–e, (f), g, j, r, (t), u,
w, z, 36. Yoxall Lodge, 1837 (Babington 1897). The typical plant
with bright orange flowers becoming blue is rare and seems to
prefer dry sandy roadsides and heathy banks. But the var *dubia*
(Arrond.) Rouy, which has creamy white flowers changing to blue
and which is very much commoner, prefers damp rather than dry
sandy soils and is often found in woodland rides and marshy
fields. (Map p 224).

M. ramosissima Rochel   Early Forget-me-not
Almost confined to the limestone in the north east and the sandstone in the south west, where it grows in dry shallow soil about
rocks and on the heaths: H–J, N–P, e, r, w, 21. 'Not rare on the
limestone' (Garner 1844). 'Sandy lanes, Trysull, Kinver' (Fraser
1864). In 1966 it was found in Brocton gravel pit on Cannock
Chase. (Map p 224).

Lithospermum officinale L.   Common Gromwell
Old records for the ruins of Tutbury Castle and Croxden Abbey
(Dickenson 1798); Grove Mill, Stafford Castle churchyard (Clifford
1817); Cheadle (Carter 1839); Burton, Alton (Garner 1844);
Horninglow (Brown 1863); Wren's Nest (Fraser 1864!); Stapenhill
(Burton Flora 1901); Oakamoor (Berrisford 1902!).

L. arvense L.   Corn Gromwell
Formerly 'not unfrequent' in cornfields (Dickenson 1798), but the
only recent record is for Burton (Harlond 1940!): (N), b, (m–n),
(r), 1.

Echium vulgare L.   Viper's-bugloss
Sandy fallow fields, particularly in the south west: (B), C, (K), L,
(N), (Y–Z), b, (d), e, (f–g), (r–t), w, 5. 'By the roadside, about a
mile from Hamstead towards Wolverhampton' (Waring 1770).
Often recorded in the past for Dudley Castle and Kinver. Recent
records for Stoke (1933!); Enville, ploughed heath (1946!);
Hollinsclough, casual in cornfield, 1947 (1950); Huntington,

luxuriant plants on a coalpit mound, SJ/970129 (1961!); Burton, abundant in a clearing in Shobnall Wood (1968).

CONVOLVULACEAE

Convolvulus arvensis L.   Field Bindweed
Arable fields and waste ground, particularly in the south: F–H, K–N, S–b, d–y, 213. Pitt, 1794. (Map p 225).

Calystegia sepium (L.) R.Br.   Hedge Bindweed
'Moist hedges' (Pitt 1796) and 'Hedges in watery places' (Dickenson 1798): A–C, F–H, K–z, 337. (Map p 225).

C. pulchra Brummitt and Heywood   Hairy Bindweed
Hedgerows near houses here and there: G, K–L, R, U, X–Y, b, e, k, p, r–s, w–x, 18. Brocton (Daltry 1923!).

C. silvatica (Kit.) Griseb.   Large Bindweed
Railway embankments, roadside hedges, rubbish tips and waste ground near houses: widespread: F–M, Q–U, W–y, 198. Cheadle (Masefield 1883!). (Map p 225).

Cuscuta epithymum (L.) L.   Dodder
Probably a fugitive introduction: no recent records. On flax and gorse (Dickenson 1798). On clover at Ford Hayes in 1870 (Garner 1871). Garner refers to this in *Science Gossip*, 1872, p 139: 'I found this plant in our county of Stafford for the first time in 1870, growing on red clover. Last year there was no clover in the field, but the plant appeared on some vetches.' King's Bromley (Moore 1897); Winshill (Burton Flora 1901); Cheadle in 1911 (1912); on *Galium saxatile* at Ramsor (1916).

SOLANACEAE

Lycium barbarum L.   Duke of Argyll's Tea-plant
Hedgerows near houses: (F), H, L, (N), U, W, (Y), Z, f–g, n, r–t, x–y, 20. Weston (Ridge 1895!).

Atropa bella-donna L.   Deadly Nightshade
Castle banks and moats, and old quarries on the Silurian limestone: local: (N), P, (Y), b, (k), (n), s, (w–x), 5. Dudley Castle (Withering 1787). 'Amongst the lime works of Sedgley and Dudley, very common' (Pitt 1794). Still present in several localities in this part of the Black Country. Other old records for Moseley near Bushbury, the fosse of Alton Castle, Stafford Castle and Tutbury Castle,

where it still persists. In 1965 it was found at Mayfield, at the foot of a stone wall near Standcliff Farm, SK/152457.

Hyoscyamus niger L.   Henbane
Recorded for waste ground in former times, but rarely seen today: (F), (L), (N), (Y), b, (k), (n), p, s, 3. Pitt, 1794. Recent records for Burton (Burges 1944); West Bromwich, garden weed (1949); Tamworth, neglected garden at Wigginton House (1965!).

Solanum dulcamara L.   Bittersweet
'In pits and hedges' (Pitt 1794): B, E–H, K–z, 627.

S. nigrum L.   Black Nightshade
Arable fields, gardens and rubbish dumps: uncommon, except perhaps near Burton and Tamworth: B, G, K, (N), (T), (Y), Z–c, f–g, n–p, (r), w–x, 24. 'Several places about Stone' (Forster in Clifford 1817).

Datura stramonium L.   Thorn-apple
Occurs sporadically in gardens and on waste ground, in some years affording a crop of records and in others none: F–G, L–N, R, T, b, e, (f), (j), n–p, s, 21. Garner, 1844.

SCROPHULARIACEAE

Verbascum thapsus L.   Great Mullein
Recorded for 'hedge banks in a sandy soil' (Dickenson 1798) and 'dry waste ground' (Brown 1863): frequent: G–N, (S), T–W, (X), Y–b, (e), f–g, j–r, (t), w–x, z, 43. Garner (1844) thought it was particularly common on limestone rocks.

V. lychnitis L.   White Mullein
Unconfirmed records for Kinver, near the Rock Houses (Stokes in Withering 1787); hedge banks near Wombourn (Wainwright in Shaw 1801); Whittington Common (Fraser 1877!); Arley Wood (Bagnall 1901).

V. nigrum L.   Dark Mullein
'Rather above a mile beyond Hamstead, towards and near Barr, Staffordshire' (Waring 1770). 'Plentiful in a lane leading from Tower Hill Farm, Perry Barr, into the old Walsall road' (Ick 1837). Near Wombourn (Garner 1844).

V. blattaria L.   Moth Mullein
'Hill Ridware, in gravelly soil' (Dickenson 1798). 'Common about

Dunsley and Kinver' (Scott in Purton 1817). Found by Mrs
Reynolds near Quinton's Orchard Farm in 1848 (Reader 1923).
Stoke, 1854 (Garner MS). Near Madeley church in 1948, G. J. V.
Bemrose (1950). These records are all uncertain. Some of the older
botanists thought that Scott mistook *V. virgatum* for *V. blattaria*.

V. virgatum Stokes   Twiggy Mullein
Grass field between Seisdon and Fox Inn (Fraser 1886!).

Misopates orontium (L.) Raf.   Lesser Snapdragon
Recorded, no doubt always as a casual, for Himley (Bree in Purton
1817); Burton (Garner 1844 and Curtis 1930); Oakamoor
(Berrisford 1907!); Pound Green Common near Upper Arley
(1959!).

Antirrhinum majus L.   Snapdragon
'Abounding on the walls of Rushall Castle, near Walsall' (Pitt
1794); Burton Abbey (Garner 1844); waste places by canal,
Whittington, Stewponey, Kinver (Fraser 1878!); Stafford (Moore
1889!); railway at Burton (BEC 1926 Rep).

Linaria purpurea (L.) Mill.   Purple Toadflax
'It grew for many years on the old abbey wall near Burton church'
(Brown 1863); Brindley Heath, a single plant by the roadside
leading to Penkridge Bank (1950).

L. repens (L.) Mill.   Pale Toadflax
Great Haywood (Fraser 1873! and Moore 1889!); Hamstead
railway cutting (Bagnall 1901); Burton (Burges 1944); Ipstones,
down steep hill into Froghall (1952!); Pelsall, railway embankment
at bridge, SK/032046 (1957!); Newcastle Lane, Stoke (1964).

L. vulgaris Mill.   Common Toadflax
Borders of cultivated fields, roadsides, railway banks, pit mounds
and waste ground, particularly in the towns: abundant in the
Black Country: B–C, F–N, Q–b, d–n, r–z, 261. 'Common in
hedges' Pitt 1794). (Map p 225).

Chaenorhinum minus (L.) Lange   Small Toadflax
Cornfields in former days, but nearly all the recent records are for
railway tracks: B, G–H, K, M–P, S, U–V, Y–Z, b, (d), f, p, (t), y, 42.
'Cornfields, rare; Blymhill amongst wheat in the Pit-down'
(Dickenson 1798). (Map p 226).

Kickxia elatine (L.) Dumort. Sharp-leaved Fluellen
Field near Highgate Common (Fraser 1865!); near Stoke, Garner (1884); cornfield, King's Bromley Hays (Reader 1922!).

Cymbalaria muralis Gaertn., Mey. and Scherb. Ivy-leaved Toadflax
Old walls: frequent: B–C, (F), G–J, L, N–Q, S–U, W–Z, b–g, j–k, n, r–s, v–w, y–z, 63. Walls at Lapley (Pitt 1817).

Scrophularia nodosa L. Common Figwort
Damp woods and ditch banks: B–g, i–w, y–z, 396. Pitt, 1794.

S. auriculata L. Water Figwort
River and canal sides: common in the south: K, (M), P–S, U–x, 184. Pitt, 1794. (Map p 226).

Mimulus guttatus DC. Monkeyflower
On river shingle and in shallow streams: frequent all down the eastern side of Staffordshire, but rare in the west: C–D, (G), H–J, (L), N–P, T–V, Y–a, f, n, r, t, (w), 36. First seen in Staffordshire by the North Staffordshire Field Club near Longton Hall pool (now drained) in 1868. There are several specimens in Fraser's herbarium from the neighbourhood of Kinver, 1877–83, and one from Cannock Chase, 'Brook at base of encampment of Swords' (1873!). In 1882 Goodall said it was 'doubtfully wild in Winnoth Dale near Tean.' Wild or not, it is still there. (Map p 226).

M. moschatus Dougl. ex Lindl. Musk
Churnet bed, Eastwall (Berrisford 1907!); Brindley Heath, side of stream near the waterworks (1945!); Ellastone, side of lake at Calwich Abbey (1947!).

Limosella aquatica L. Mudwort
In wet mud at the edges of ponds and reservoirs when the water has receded in a dry season: rare. Stowe Pool (Power MS undated); Knypersley Pool, 'when it was dry', 1887, and Rudyard Lake (Painter 1892); seen again at Rudyard Lake in 1934 and 1969, in the bay below the youth hostel; Harborne Reservoir, Pottal Reservoir, Hayhead (Bagnall 1901); mud of dried up pond at Slitting Mill near Rugeley (Reader 1923!); Stanley Pool, near the footbridge (1969).

Digitalis purpurea L. Foxglove
Roadsides, woods, heaths and gritstone moors: A–z, 663. Norton in the Moors with white flowers (Plot 1686).

Veronica beccabunga L. Brooklime
Shallow streams, ditches and marshy places: A–w, y–z, 551.
Yoxall Lodge (Gisborne 1787!).

V. anagallis-aquatica L. Blue Water-speedwell
Ditches and streams: apparently rare, but perhaps overlooked.
Forton, ditch near Windswell Pool (1954!); Wolfscote Dale below
Gipsy Bank (1960); Lode Mill (1960); Whiston Brook near Mitton
Manor (1961!).

V. catenata Pennell Pink Water-speedwell
Pond margins, ditches and shallow streams: common in central
Staffordshire: D, J–K, P, V–a, c–h, n–p, 42. The early botanists
did not distinguish this species from the last, but most of their
records probably refer to *Z. catenata*. That for Cow Leasow ditch,
Tixall, 'corolla of a beautiful pink' (Clifford 1817), certainly does.
(Map p 226).

V. scutellata L. Marsh Speedwell
Marshy fields and pond margins, in sandy or peaty soils: local:
(B), (G), H, (K), L, (N), R–S, U, W–Y, (Z), (b–c), d–e, (f–g), (k),
(n–p), w, (z), 14. 'Ditches about Tamworth' (Withering 1796).
Var *villosa* Schumach. has been recorded for Eccleshall, marshy
field behind the church (1953!); Enville, Foucher's Pool (1954!);
and Uttoxeter, pond at the Dearndales (1956!).

V. officinalis L. Heath Speedwell
Open heathy woods, dry banks, sandy commons and limestone
hills: common in the Churnet Valley and moorlands: B–C, G–P,
R–U, W–X, (Y), Z–a, (b), d–f, j, (k), n, r–s, v–w, z, 103. Yoxall
Lodge (Gisborne 1787!). (Map p 227).

V. montana L. Wood Speedwell
Moist loamy woods: A–B, E–U, W–g, i–j, n, r, v–w, z, 161. Weston-
under-Lizard (Dickenson 1798). (Map p 227).

V. chamaedrys L. Germander Speedwell
Hedgebanks: B–z, 663. Yoxall Lodge (Gisborne 1792!).

V. serpyllifolia L. Thyme-leaved Speedwell
Damp meadows, cultivated fields and woodland rides: B–U, W–j,
n–r, u–w, z, 206. Yoxall Lodge (Gisborne 1787!).

V. arvensis L. Wall Speedwell
Cornfields on sandy soils, gravel pits, dry bare places on sandy

heaths, walls, railway tracks and limestone turf: B–C, E, H–P, R–U, W–f, h–k, r, u–w, y, 122. Dickenson, 1798.

V. hederifolia L.   Ivy-leaved Speedwell
Moist shady places in hedgerows and copses, gardens, arable fields and in grass at the foot of limestone cliffs: B, H–L, (N), P–U, W–Z, b–e, (f), g, j–n, r–w, y, 110. 'Sometimes very much abounding amongst wheat very early in the spring' (Pitt 1794).

V. persica Poir.   Common Field-speedwell
Now a common weed in cultivated fields: C, (F), G–P, R–w, y, 286. Trysull (Fraser 1864!).

V. polita Fr.   Grey Field-speedwell
A rare garden weed. Hawkesyard (Reader 1922!); Madeley (1945!); Oakamoor in 1952; Thorpe Constantine (1957!); Seighford (1958!). Earlier records are uncertain.

V. agrestis L.   Green Field-speedwell
Gardens, waste ground and cultivated fields: frequent: A, C, G–P, R–U, W–b, e, (f), g–k, n–q, (r), s–w, y–z, 75. Dickenson, 1798.

V. filiformis Sm.   Slender Speedwell
Churchyards, lawns, canal sides, road verges and river banks: frequent and increasing: B, H–J, N, S, W–a, j, r, v–w, 19. Canal side at Sutton (1955!). Abundant in the churchyard at Ilam (1970).

Pedicularis palustris L.   Marsh Lousewort
Wet places on heaths and moors and over limestone: local: C, (G), (K–L), (N), P, R–S, (W), (Y–Z), (g), (n), 8. 'Wet meadows' (Dickenson 1798). Recent records for the Downs Banks, Stone (1942!); Moss Carr (1946!); Burnt Wood (1947); pond near Calton, SK/115493 (1958); and others, unpublished, for Fawfieldhead, SK/060611; Offleybrook, SJ/785304; Podmore Pool; and Stanton, SK/1247.

P. sylvatica L.   Lousewort
Acid bogs on heaths and moors: frequent: B–C, G–P, R–U, W, (X–Y), Z–a, (b), d, f, (n), r–s, (t), (w), z, 58. Needwood Forest (Pitt 1794). (Map p 227).

Rhinanthus serotinus (Schonh.) Oborny   Greater Yellow-rattle
Unconfirmed and doubtful records for Ashley and Wetley (Garner 1844); and near Stafford (Douglas 1851).

R. minor L.   Yellow-rattle
Hayfields and dry roadsides, but sometimes in damp grassland:
common, particularly on the limestone: B–P, R–U, W–b, d, f–g,
j–x, z, 180. Pitt, 1794. All the local specimens in my herbarium
belong to subsp *minor.*

Melampyrum pratense L.   Common Cow-wheat
Heathy woods on acid rock: local: (B), C, F–H, (K), (M), N, R,
(S–T), (Z), (b), (d), e, (g), t, (w–x), z, 16. Pitt, 1796. There are
thriving colonies between Cheadle and Alton and in the woods
north and south of Loggerheads.

Euphrasia officinalis L. sensu lato   Eyebright
Common in hayfields and by roadsides in the north east of the
county: B–D, (F), G–J, (K), L–P, (R), S–U, W–X, (Y), (b), e, (g),
k–m, (s–t), (w), 96. Pitt, 1794. Samples from 130 sheets of Staffordshire specimens were submitted to Dr P. F. Yeo in 1963 and the
following account of the segregates rests upon his determinations.
But acknowledgement is also due to H. W. Pugsley who named
many of the earlier specimens. (Map p 227).

E. nemorosa (Pers.) Wallr.
This species has a wider range in Staffordshire than any of the
others and grows more frequently in artificial habitats: C, H–J, L,
N, e, k, 12. Pure *E. nemorosa* or hybrid plants which have a pro-
nounced strain of *E. nemorosa* have been found on coalpit mounds
at Silverdale, by the locks of canals, in several places on Cannock
Chase and abundantly there in gravelly soil between the cement
foundations of a wartime camp. In the north it grows on heathy
banks and by roadsides on the limestone. The specimens Pugsley
saw came from the north of the county and were all attributed to
var *collina* Pugsl. But some of these are now thought to be crossed
with *E. confusa.*

E. confusa Pugsl.
Local in dry upland pastures and by roadsides in hilly districts:
B–C, G–J, N–P, 27. There are fine examples on the Roches at about
1,100ft and in many places on the limestone. *E. confusa* x *nemorosa*
has been recorded fourteen times.

E. borealis (Townsend) Wettst.
Hayfields and grassy roadsides in the north: locally plentiful:
C–D, G–J, L, N–P, T, 25. This is the correct name (Yeo 1970) for

the taxon we used to call *E. brevipila*. We have two forms, one with short glands and one almost entirely without glands. Plants which appear to be *E. borealis* x *confusa* have been recognised in seven places.

E. rostkoviana Hayne
Local in upland hayfields, especially in the extreme north, as about Flash: B–C, H, N–P, 8.

E. anglica Pugsl.
Local in damp grassland in the north east: B–D, G–J, N–P, 15.

Odontites verna (Bellardi) Dumort.   Red Bartsia
Rough grassy places by roadsides and at field entrances: frequent, especially in the centre of the county: H, M, P, S–e, g, j–m, r, t, v–w, 75. Pitt, 1794. Dickenson (1798) described it as common in pastures and cornfields in a moist soil. Most of our plants are probably best placed under subsp *serotina* (Wettst.) E. F. Warb.

OROBANCHACEAE

Lathraea squamaria L.   Toothwort
Frequently seen on the roots of wych elm and hazel in the Manifold Valley, Dydon Wood and Forest Banks in the north east and about Dudley Castle in the south: rare elsewhere: H–J, M–P, S, U, (V), a, s, w, 17. 'Longley meadow in King's Bromley, and by the side of Yoxall brook' (Riley in Shaw 1801). Often plentiful at Ilam along Paradise Walk.

Orobanche rapum-genistae Thuill.   Greater Broomrape
Unconfirmed records for 'Blymhill, in the Pye-hill Lane, near Gorse' (Dickenson 1798); Dimminsdale (Carter 1839) and later records for Cheadle (1890) and the Ranger (Berrisford 1904!); Heleigh Castle (Garner 1844); near Stafford (Douglas 1851); on broom at Keele (1903). The specimen from the Ranger is attached to a piece of broom.

LENTIBULARIACEAE

Pinguicula vulgaris L.   Common Butterwort
Plentiful in one small bog on Cannock Chase (1945!), where *Parnassia palustris* flowers later in the year: rare or extinct in its other recorded stations. 'In the grounds adjoining Mr Jervis' Pool near Yarnfield' (Forster 1796); Belmont (Sneyd 1796); 'Near

Blymhill, in Mrs Brew's moor, in a ditch' (Dickenson 1798); 'A boggy valley on Cannock Heath, nearly opposite to Tixall gate, 1811' (Clifford 1817); Whittington Heath (Power MS undated); near Stanton (1930 and 1938!); in a bog at Croft Bottom near Hollinsclough (1951).

Utricularia vulgaris L. sensu lato   Greater Bladderwort
Disused canals, ponds and deep drains: rare: K, (L), (R), (W), Y, (Z), (d), (g), (n), 2. Ditches near Wychnor (Jackson 1837). Shelmore Wood (Bagnall 1901), an undated specimen is labelled *U. neglecta* Lehm. Pond by railway between Rugeley and Colwich (Perry 1924!); drain near Betley Mere (1955!); canal at Great Haywood.

U. minor L.   Lesser Bladderwort
Unconfirmed and uncertain records for Chartley Moss and Norton Bog (Bagot in Withering 1801); near Betley (Purton 1821); Craddocks Moss (Garner 1844).

VERBENACEAE

Verbena officinalis L.   Vervain
'Waste places about villages' (Dickenson 1798): rare now, formerly 'not unfrequent' (Dickenson 1798): (N), W, (Y–Z), (b), (n), 2. Old records for Lichfield, Tutbury, Stowe, Tatenhill, Winshill, Hopton and Oakamoor: recent ones for Shebdon (1943) and Forton (1944!).

LABIATAE

Mentha pulegium L.   Pennyroyal
'Plentiful on commons on the Cheshire borders of Staffordshire' (Garner 1844). If this was ever true, it is not true now. Blymhill Heath (Dickenson 1798); bank of Pensnett Reservoir (Scott 1832); Lichfield (Jackson 1837); Craddocks Moss (Garner 1844); Seckley Wood (Fraser 1874!); mud flats at head of Rudyard Lake (Allen 1911!).

M. arvensis L.   Corn Mint
'In wet cultivated land, frequent in stubbles' (Dickenson 1798). This is probably still true, though field records have not been kept. Recent specimens from twelve scattered localities, most of them from arable fields, but some from the wet margins of ponds.

M. x *gentilis* L. (*M. arvensis* x *spicata*)
'Lane above Bentley Farm near Blithbury' (Reader 1920!);

Ellastone, lakeside at Calwich Abbey (1947!); Meerbrook (1950!); canal at Bucknall (1952!); Moss Lane near Cheadle (1958!).

M. aquatica L.   Water Mint
Marshy places by ponds, rivers and canals: B–C, E–z, 355. 'Sides of the river at Tamworth' (Withering 1796).

M. x *verticillata* L. (*M. aquatica* x *arvensis*)
Wet places by rivers, ponds and canals: widespread and probably common, though field records have not been kept. Oakamoor (Carter 1839 and Berrisford 1902!).

M. x *piperita* L. (*M. aquatica* x *spicata*)
Damp roadsides, often near houses: widespread, but uncommon. 'Near the river, Tamworth' (Withering 1796). Lane between Wergs and Codsall (Fraser 1873!); Cheadle (Masefield 1883!); Oakamoor (Berrisford 1905!); Marchington (Ridge 1916!); Tixall (Thornton 1923!); Weston-on-Trent (Reader 1923!); Silverdale (1943!); Dimminsdale, Oakamoor (1944!); Solomon's Hollow near Leek (1950!); Bagnall (1955!).

M. spicata L.   Spear Mint
Garden escape. Riversides and roadsides. Near Cheadle (Bourne in Dickenson 1798); under Heleigh Castle bank (Garner 1844); Trent side, Wychnor (Brown 1863); roadside between Swindon and Highgate Common (Fraser 1885!); Patshull (BEC 1922 Rep); Okeover, river bank (1950).

M. longifolia (L.) Huds.   Horse Mint
Roadsides and waste ground. The few Staffordshire examples I have seen resemble the illustration in Ross-Craig's *Drawings of British Plants* and have included stamens. Garner (1844) said it could be found occasionally. Purchas (1885) saw it in very small quantity in one place in Dovedale and Reader found it at Longdon (Bagnall 1901). Recent records for Ipstones Edge (1952); roadside near Highgate Common (1953!); Fawfieldhead (1955!); Eccleshall, waste ground on site of war factory (1961!).

M. x *niliaca* Juss. ex Jacq. (*M. longifolia* x *rotundifolia*)
Garden escape. Riversides and waste ground. Dimminsdale (Berrisford 1903!); Mayfield, riverside (1947!); Oakamoor (1952!); Huntley near Cheadle (1955!); roadside at the Red Cow inn near Uttoxeter (1956!); near Elm Cottage, Brewood (1957!). There are

(*above*) Chartley Moss, home of the Round-leaved Sundew (*Drosera rotundifolia*), Cranberry (*Vaccinium oxycoccos*) and Bog Rosemary (*Andromeda polifolia*); (*below*) Loynton Moss, the reed bed (once Blakemere Pool), where the Greater Spearwort (*Ranunculus lingua*) grows

(*above*) Coombes Valley; (*below*) the river Manifold at Ilam

no old records under this name, but the few we have of *M. rotundifolia* (L.) Huds. should probably be transferred here. A specimen collected at Croxden (Allen 1910!) certainly should. This and all the other Staffordshire plants I have seen belong to var *alopecuroides* (Hull) Briq.

Lycopus europaeus L.   Gipsywort
'Moist places, by the side of pools, frequent' (Dickenson 1798): particularly common along the canals: E–G, K–M, (N), Q–u, w–y, 228. (Map p 228).

Origanum vulgare L.   Marjoram
Common on rocky limestone hills and occasionally elsewhere on old ruins and road verges: C, H–J, (K), (N–P), T, (U), b, e, g, s, 14. Bunster Hill (Pitt 1794).

Thymus pulegioides L.   Large Thyme
Near Whittington Heath (Fraser 1877!).

T. drucei Ronn.   Wild Thyme
Common on the Carboniferous limestone and about Dudley, Kinver and Arley: C, H–J, N–P, (Y), (e–f), r–s, v–w, z, 38. 'Heaths and roadsides in the south part of the county' (Pitt 1794). (Map p 228).

Calamintha ascendens Jord.   Common Calamint
'Hedge banks near Wolseley Bridge' (Dickenson 1798); Dudley Castle (Shaw 1801); Lichfield, Heleigh Castle, Tutbury Castle (Garner 1844); Stretton (Brown 1863); Heleigh Castle (Fraser 1869!); Hopton (Bagnall 1901). Ridge (Flora) claimed to have seen it in the Manifold Valley and Dovedale, but he may have been mistaken. There are no recent records.

Acinos arvensis (Lam.) Dandy   Basil-thyme
About limestone outcrops in the Manifold Valley and Dovedale and formerly in sandy fields in the south west: H–J, (V), (f), (w), 6. Lichfield (Jackson 1837). There are several records for the limestone and also the following: Kinver, with white flowers (Garner 1844); Whittington Heath (Fraser 1877!); railway embankment at Rocester (1907); stubble field near Armitage (Reader 1919!).

Clinopodium vulgare L.   Wild Basil
Dry hedgebanks and busy hillsides, particularly on the limestone and in the south west: (G), H–J, (N), P, R, (W), (Y–Z), b, d,

(n–p), r–s, v–w, z, 27. Dudley Castle and near Tamworth Castle (Withering 1796). (Map p 228).

Melissa officinalis L.   Balm
'Plentiful about Kinfare, Staffordshire' (Brunton in Withering common (BEC 1923 Rep).

Salvia verticillata L.   Whorled Clary
Burton (Curtis 1930); North Street, Stoke (1932!).

S. horminoides Pourr.   Wild Clary
'Plentiful about Kinfare, Staffordshire' (Brunton in Withering 1787); Castle Hill, Tamworth (Withering 1796); Tutbury Castle ruins (Dickenson 1798); Whitmore (Garner MS); Oakamoor station (Berrisford MS); near Wednesbury in 1951.

Prunella vulgaris L.   Selfheal
'Meadows and pastures' (Pitt 1796) and woodland rides and clearings: A–z, 589.

Betonica officinalis L.   Betony
Rough gorsy pastures and bushy hillsides: frequent and locally common: A, C–b, d–g, j–n, r–z, 171. Pitt, 1794. (Map p 288).

Stachys arvensis (L.) L.   Field Woundwort
Frequent in sandy arable fields: K, R–S, W–X, (b), c, e–g, j–r, t–w, y, 24. Dickenson, 1798.

S. palustris L.   Marsh Woundwort
'Moist places and banks of rivers' (Dickenson 1798): particularly common by canal sides: B, G–Q, S–T, W–Y, b–w, y, 100. (Map p 229).

S. x *ambigua* Sm. (*S. palustris* x *sylvatica*)
Wet lanes and canal sides and also rubbish tips, manure heaps and waste ground near farm buildings: frequent: F, G, (H), L–M, (N), T, W–Y, (b), d, g, (j), 15. Garner, 1844.

S. sylvatica L.   Hedge Woundwort
Woods and hedgerows: A–z, 738. Pitt, 1794.

Ballota nigra L.   Black Horehound
Common on waste ground in the south, especially near houses: K, (M–N), R–U, W–x, z, 169. Stafford (Stokes in Withering 1787).

Lamiastrum galeobdolon (L.) Ehrend. and Polatsch.

Yellow Archangel

Common in our richer woods, chiefly on limestone and marl:
A–B, E–F, H–R, T–U, W–g, j–n, r–z, 181. Withering, 1787.
(Map p 229).

Lamium amplexicaule L.   Henbit Dead-nettle
Cultivated sandy fields: frequent and locally common: (G), K,
(N), R–T, X, (Y), Z, (b), d–g, k, n–s, w–y, 42. Dickenson, 1798.

L. hybridum Vill.   Cut-leaved Dead-nettle
Said to be frequent (Garner 1844), but we cannot be sure that it
was always correctly understood. Turnip field near Armitage,
'corolla tube internally naked' (Reader 1921!); Chebsey, hedgebank
in a lane, SJ/857292 (1958!); Mickle Hills near Lichfield, edge of
cornfield at top of sandy roadside embankment (1962!).

L. purpureum L.   Red Dead-nettle
Cultivated and waste land: often growing with *Stellaria media* in
neglected gardens in moist soil: B, E–z, 435. Pitt, 1794.

L. album L.   White Dead-nettle
Roadsides and waste ground, especially near houses: B–z, 634.
Pitt, 1794.

L. maculatum L.   Spotted Dead-nettle
Hedgebanks in villages: B, E, (G), H–J, L–P, (R–S), T–U, Y–a,
(b), c, (t), w, 20. Burton (Brown in Garner 1844).

Leonurus cardiaca L.   Motherwort
Gornal Wood (Wainwright in Shaw 1801); Iverley Hills and
adjoining fields (Scott 1832); 'In a narrow shady lane, among
nettles, at the back of Perry Barr Park' (Ick 1837); garden at
Oakamoor (Berrisford 1902!); Great Haywood in 1907 (1916) and
again in 1924 (Thornton!); once near the canal at Trentham (Ridge
in Flora); West Bromwich (1948).

Galeopsis angustifolia Ehrh. ex Hoffm.   Red Hemp-nettle
'In a bean-field near the toll-bar between Stone and Stafford, 1839
and 1841' (Garner 1844); between Wychnor and Barton, 1845
(Garner MS); known for many years in an old lime quarry or
quarries at Sedgley Beacon (Fraser 1864! and Miss Bigwood
1952!).

G. tetrahit L.   Common Hemp-nettle
Arable and waste land, roadsides and clearings in woods: B–z, 535.
'Cornfields and gardens' (Pitt 1796).

G. speciosa Mill.   Large-flowered Hemp-nettle
Roadsides and cultivated fields, with a preference for moist peaty
soil: sometimes abundant, as in Keele Park in 1956, but as a rule
only a few plants are seen in any one place: C, F, (H), K–L, (M),
R, (S), T, (Y–Z), a, (b), d–e, (f–g), h, (j), y, 14. Blymhill (Dickenson
1798). 'Cornfields near Newcastle' (Pitt 1817). A third of the
recent records are for the countryside near Newcastle.

Nepeta cataria L.   Catmint
Dry roadsides, waste ground and castle ruins: rare, except perhaps
on the limestone in the extreme south. Dudley Castle (Withering
1796); Croxden Abbey (Dickenson 1798); 'Near the farmyard,
Tower Hill, Perry Barr' (Ick 1837 and Bagnall 1882!); Heleigh
Castle and near Stourbridge (Garner 1844); Tutbury Castle
(Brown 1863 and Nowers 1888!); lane between Trysull and Swindon
(Fraser 1865!); hedge between Seisdon and Beech House (Fraser
1878!); canal side at Wordsley (1959).

Glechoma hederacea L.   Ground-ivy
Woods and hedgebanks: A–z, 519. Near Yoxall Lodge (Gisborne
1792!). (Map p 229).

Marrubium vulgare L.   White Horehound
Dickenson (1798) said it was frequent by roadsides, but there are
few records and no very recent ones. Iverley (Scott 1832); Lichfield
(Jackson 1837); Pound Green near Arley (Fraser 1873!); garden,
Oakamoor (Berrisford 1902!); near old kitchen garden, Beaudesert
Park (Reader 1920!); Burton (Curtis 1930).

Scutellaria galericulata L.   Skullcap
Ditches and the margins of ponds, lakes and canals: B, E, (F),
G–H, K–U, W–u, w–y, 152. Pitt, 1796. (Map p 229).

S. minor Huds.   Lesser Skullcap
Wet heaths: rare. Brakenhurst Bog (Gisborne 1791!); Bagot's Park
(Withering 1801 and Masefield 1885!); Norton Bog (Withering
1801); near Swynnerton (Purton 1821); Enville Common (Fraser
1865!); Foucher's Pool (Fraser 1873!); Seckley Wood (Fraser
1884!); Sherbrook Valley, Cannock Chase (Bagnall 1901): Consall
(1902); Penn Common, Miss Hibbert (1967!).

Teucrium scorodonia L.   Wood-sage
Dry open habitats on sandstone and limestone, hillsides, hedge-
banks, wood clearings and gravel pits: A–U, W–g, j–z, 339.
Pitt, 1794. (Map p 230).

Ajuga reptans L.   Bugle
Meadows and moist woods: B–r, t–x, z, 319. Pitt, 1794.

PLANTAGINACEAE

Plantago major L.   Greater Plantain
Roadsides, waste ground, tracks and paths: A–z, 792. Near Yoxall
Lodge (Gisborne 1791!).

P. media L.   Hoary Plantain
Common in limestone pastures: H–J, N–P, (S), W, a–b, (f–g), h,
m, p, s–t, w–x, 56. 'At the Temple' (Gisborne 1791!). (Map p 230).

P. lanceolata L.   Ribwort Plantain
Fields, roadsides and waste ground: A–z, 793. Pitt, 1794.

P. maritima L.   Sea Plantain
Remains of Tixall salt marsh (Thornton 1923!), where Reader
(1924) said it was fairly plentiful; Pasturefields, old salt marsh
between the canal and the river (1947! and 1956!).

P. coronopus L.   Buckshorn Plantain
Dry bare places on sandstone heaths: local: (L), (N), R, Y, (b),
e–f, (g), m, r, w, 11. 'Lanes in a gravelly soil, not unfrequent'
(Dickenson 1798). Carter (1839) found it sparingly near Cheadle,
Garner (1844) in sandy places at Trentham, Brown (1863) at
Branston and Barton, and Reader (1918!) on gravel paths at
Hawkesyard. But most of the records, old and new, are for Cannock
Chase and the south west.

P. indica L.
Casual. North Street, Stoke, G. J. V. Bemrose (1932!); Burton
(BEC 1938 Rep).

Littorella uniflora (L.) Aschers.   Shoreweed
Lakes and reservoirs, in wet sand at the edge of the water: rare.
Pensnett Reservoir (Scott 1832); Trentham Pool and Calf Heath
(Garner 1844); Hednesford Pool (Garner 1844 and Fraser 1865!);
Knypersley Pool in 1887, when it was dry (Painter 1892); Gailey
Reservoir (Bagnall 1901 and Thornton 1923!); Rudyard Lake, at
the northern end (Garner 1844 and Edees 1933!).

CAMPANULACEAE

Wahlenbergia hederacea (L.) Reichb.   Ivy-leaved Bellflower
'Betwixt Rugeley and Beaudesert' (Power MS undated). This is
probably the source of Jackson's record for near Lichfield (1837)
and of Brown's statement (1863) that it grew in wet places on
Cannock Chase. The only recent record is for Biddulph Grange,
where it was found in grass in wet peaty soil at the edge of the lily
pond in front of the house (1967!). In such a situation we cannot be
sure that it was not introduced.

Campanula latifolia L.   Giant Bellflower
In woods in nutrient-rich soils, chiefly on the limestone, though not
confined to it: local: D, (F), H–J, (L), M–P, S–U, (V), X, (Y),
Z–b, d, g, i, (n), (r), v, (x), 51. 'In the mountainous parts of . . .
Staffordshire . . . plentifully' (Ray 1670). Forster (1796) found it
with white flowers in the lane leading from Meaford Farm to
Meaford. Garner (1844) said the white-flowered plants were still
there in 1836. (Map p 230).

C. trachelium L.   Nettle-leaved Bellflower
Woods and shady hedgebanks on limestone and marl: today less
frequently recorded than *C. latifolia,* though the early botanists
found it locally plentiful: H, (N), R, (V), X, (Z–a), b, (g), j, (n),
(s–t), v, (x), z, 10. 'Yoxall, Wood Lane, abundantly, 1792' (Gisborne
MS). 'Plentiful in the shady lanes between Perry Barr and Great
Barr' (Ick 1837).

C. rapunculoides L.   Creeping Bellflower
Garden escape. 'It grows wild in hedgebanks at Leigh, near
Uttoxeter' (Smith 1871); hedge near Tamworth (Bagnall 1901);
Shenstone (Harlond 1939!).

C. glomerata L.   Clustered Bellflower
Dudley Castle hill and Rowley lanes (Scott 1832); Haughton
(Moore 1889!), probably a garden escape .

C. rotundifolia L.   Harebell
Grassy places in light, usually dry shallow soils: B–z, 399. Stone
(Forster 1796).

C. patula L.   Spreading Bellflower
Old records, but no recent ones, for hedgebanks and woods,
particularly in the neighbourhood of Lichfield. 'Near the bath at

Lichfield' (Woodward in Withering 1787); Barton Lane (Gisborne MS), where Babington saw it in 1832 (A.M.B. 1897); between Stewponey and Stourbridge (Purton 1821); Hopwas Wood (Power MS); Burton (Garner 1844); Trysull (Fraser 1864!); Stapenhill (Burton Flora 1901); Froghall (Berrisford! undated).

C. rapunculus L. Rampion Bellflower
Enville (Withering 1796 and Fraser 1878!), abundant there in 1821 (Purton), another specimen, Fraser 1879, det Druce (BEC 1924 Rep). Other records for Dudley (Watson 1835); Blymhill (Garner 1844); Stapenhill (Brown 1863); Tamworth (Bagnall 1901); Lichfield (1911); but we do not know if they are reliable.

Legousia hybrida (L.) Delarb. Venus's-looking-glass
'In a turnip field on Kingston Hill near Stafford' (Garner 1844); Hamstead railway cutting in 1868 (Bagnall 1901); weed at Barlaston in 1871 (1905); Burton in 1971.

Phyteuma spicatum L. Spiked Rampion
Four Ashes Hall, 'under a copper beech at the edge of a big lawn next a shrubbery,' presumably introduced, but long established, Mrs Amphlett (Wilmott 1942).

Jasione montana L. Sheepsbit
Heathy banks on sandstone: local: (B), C, (F), G, (H), (K), L–N, R, (f), r, (t), z, 11. Dickenson, 1798. Perhaps most plentiful in the Churnet Valley, where Carter (1839) said it was 'common and abundant.'

RUBIACEAE

Sherardia arvensis L. Field Madder
'Cornfields, in a light soil, common' (Dickenson 1798). This is the verdict of most of the early botanists, but today it is common only on the limestone and in the south west: H–J, (K), N, S, W, (b), d–e, (f–g), k, n, r, (w), 14. Pitt, 1796.

Cruciata laevipes Opiz Crosswort
Hedgebanks, grassy roadsides and wood borders, with a preference for calcareous soils: B–D, G–g, j–n, r–w, z, 324. Near Yoxall Lodge (Gisborne 1792!). (Map p 230).

Galium odoratum (L.) Scop. Woodruff
Frequently seen outside gardens as an escape from cultivation, but

native and common in woods on the limestone and marl: B, E–K, (L–M), N–P, R–U, X–d, (g), j, n, (r), (t), v–w, 58. Needwood Forest (Gisborne 1792!).

G. mollugo L. Hedge Bedstraw
Plentiful in hedgerows in the extreme south of the county about Dudley, Kinver and Arley, but rare and fugitive elsewhere: L, Q, (a), (f), n, r–s, (t), v–w, (x), y–z, 29. Dickenson, 1798. (Map p 231).

G. verum L. Lady's Bedstraw
Dry grassy banks and pastures, especially in light soils: H–J, (K), L–U, W–z, 312. Pitt, 1796.

G. saxatile L. Heath Bedstraw
Heaths and moors: A–U, W–a, c–w, y–z, 406. Dudley Wood (Withering 1787). (Map p 231).

G. sterneri Ehrend. Limestone Bedstraw
Restricted to the rocky hillsides of the limestone dales: H–J, P, 14. Garner, 1844.

G. palustre L. Common Marsh-bedstraw
Marshy places in meadows, by canals and lakes, and on heaths and moors: A–w, y–z, 538. Pitt, 1794.

G. uliginosum L. Fen Bedstraw
Wet rushy meadows and acid bogs: often growing with *G. palustre*, but much less common: C, G–H, Q–S, W–Z, c–f, (j), m–n, r, 32. Pitt, 1794.

G. aparine L. Cleavers
Hedgerows, wood borders, arable fields and waste ground: A–z, 705. Pitt, 1794.

G. spurium L. False Cleavers
Allotments at Burton (Burges, Hardaker, Lousley and Thomas in BEC 1936 Rep), where it is now established.

CAPRIFOLIACEAE

Sambucus ebulus L. Dwarf Elder
Always near houses and probably a relic of ancient cultivation, but long established in a few places. 'About a mile from Hamstead towards Wolverhampton, sparingly; on Tamworth Castle bank, plentifully' (Waring 1770). 'Tutbury Castle bank, on the west

side, in great plenty' (Pitt 1794). There are many later records for Tutbury Castle where it still survives (1956!). 'In the turnpike road leading from Wolseley Bridge to Rugeley near the latter place' (Forster 1796). Other records for Branston (Dickenson 1798); lane below Tixall House (Clifford 1817); canal bank at Delph near Brierley (Scott 1832); near Newcastle (Garner 1844); waste ground at Fenton, 1848 (Garner MS); near the mill, Mavesyn Ridware (Bagnall 1901 and Reader 1917!); near Chartley Castle (Harlond 1935!).

S. nigra L. Elder
Hedges and thickets: A–z, 784. Plot, 1686.

Viburnum opulus L. Guelder-rose
Hedges and woods, often in damp places, throughout the county: B–C, E–z, 355. Pitt, 1794.

Symphoricarpos rivularis Suksd. Snowberry
Often seen in hedges near gardens: B–D, G–P, R–g, j–y, 158. Waterhouses (1947!).

Lonicera xylosteum L. Fly Honeysuckle
Introduced. Needwood Forest (Hewgill in Garner 1844); Sinai Park (Brown 1863); Trysull Dingle (Fraser 1864!); Chillington woods (Fraser 1878! and Miss Walker 1958!); Oakamoor (Masefield 1883!); near Knypersley Hall (Painter 1892); Belmont woods (Ridge in Flora).

L. periclymenum L. Honeysuckle
Woods and hedgerows: A–C, E–z, 528. Plot, 1686.

L. caprifolium L. Perfoliate Honeysuckle
'Gardens and plantations, but introduced' (Brown 1863). 'In an out-of-the-way place near a country wayside on the confines of Derbyshire, discovered in 1879 and seen again in the two following years' (Goodall 1882); Sinai Park (Burton Flora 1901); Alton station (Berrisford! undated); between Cheadle and Oakamoor (Ridge in Flora).

ADOXACEAE

Adoxa moschatellina L. Moschatel
Hedge bottoms and moist shady places in woods: A–g, j–n, q–s, (t), v–w, z, 217. Between Wolverhampton and Penkridge (Stokes

in Withering 1787). The earliest references to this plant are all interesting. 'April 1791, in the dingle near the parsonage, Hanbury: Callingwood Lane, near bottom of hill, right hand bank (at the top) as you go to Burton' (Gisborne MS). 'Ditch banks on this farm' (Pitt 1794). 'Woods, damp rocks and sides of rivulets' (Dickenson 1798). 'Under trees in wet ground behind Tixall churchyard' (Clifford 1817). 'Much more abundant in many places in this neighbourhood (Dimminsdale) than I ever observed it elsewhere; it abounds on every moist bushy bank' (Carter 1839). (Map p 231).

VALERIANACEAE

Valerianella locusta (L.) Betcke   Common Cornsalad
Recorded for sandy arable fields, roadside and railway banks, canal sides, walls and limestone rocks: widespread but uncommon: H, (N), P–R, X, (Y), Z, (b), (f), (n), p, (r), w, 10. 'Cornfields, frequent' (Dickenson 1798).

V. carinata Lois.   Keeled-fruited Cornsalad
Rocky limestone hillsides: plentiful about Wetton Mill: C, H–J, 5. A. Ley in Watson, 1883.

V. rimosa Bast.   Broad-fruited Cornsalad
Potato field at Mucklestone (1951!).

V. eriocarpa Desv.   Hairy-fruited Cornsalad
Unconfirmed records for Cheadle and Lichfield (Garner 1844).

V. dentata (L.) Poll.   Narrow-fruited Cornsalad
Lichfield (Jackson 1837); Cheadle (Carter 1839); Wootton, Beeston Tor, Shirleywich (Garner 1844); Stapenhill and Shobnall (Brown 1863); near Highgate Common (Fraser 1865!); Wetton (1870); allotments near Colwich and cornfields at Armitage (Bagnall 1901); fields near Bishop's Wood and at Ashley (Ridge in Flora); Sugarloaf, Wetton (Daltry 1946!).

Valeriana officinalis L.   Common Valerian
Ditches, canal sides, marshy woods: A–r, t–w, z, 439. Pitt, 1794. The taxon with three to four pairs of broad leaflets, which we used to call *V. sambucifolia*, is the common plant in Staffordshire. But a plant with more numerous and narrower leaflets occurs in drier bushy places on the limestone. Withering (1776) tells us that cats are 'delighted' with the roots.

V. pyrenaica L. Pyrenean Valerian

Known for many years in Star Wood, Oakamoor, where it can still be seen. 'I was also highly delighted to discover, by the side of a stream which runs along the bosom of the valley, *Valeriana pyrenaica*, growing in tolerable abundance. It is, I am aware, considered as one of our certainly non-indigenous plants, but no one who has seen it in the locality here given will hesitate to pronounce it decidedly wild or at least perfectly naturalised' (Carter 1839).

V. dioica L. Marsh Valerian

'Moist meadows' (Dickenson 1798): frequent, especially in the north: B–C, (F), G–P, R–S, U, W–X, (Y), Z–a, c–g, n, r, v, y–z, 68. (Map p 231).

Centranthus ruber (L.) DC. Red Valerian

Occasionally seen on old walls. Burton (Garner 1844); Oakamoor (Berrisford 1918!); Wychnor, on the walls of the church (1956).

DIPSACACEAE

Dipsacus fullonum L. Teasel

Riversides between Burton and Tamworth on the Keuper marl: locally plentiful: elsewhere odd plants are found by roadsides, especially in recently excavated drains: L, N, R–S, V, (X–Y), Z, (a), b, d, f–k, (m–n), p–r, (t), v, x–y, 30. 'Near Tamworth, common' (Pitt 1794).

D. pilosus L. Small Teasel

Woods on limestone and marl: local and decreasing, though it has persisted for many years in a few places: H–J, (P), (T), (V), (X–Y), b, (d), e, (g), n, (w), 5. Near Ranton Abbey (Gisborne 1791!) and by the gate at Burton Abbey (Gisborne MS). 'Near the bridge over the Trent at Walton by the foot way' (Forster 1796). Dickenson (1798) said it was frequent in hedges at Blymhill. Recent records for Hopwas Wood (Jackson 1837 and 1945!); Ilam, at the foot of Bunster Hill (1925 and 1946!); Waterhouses, north east slope of Soles Hill (1947!); Penkridge, streamside at Congreve (1954!); Branston, Easthill Wood (1956!).

Knautia arvensis (L.) Coult. Field Scabious

Hedgebanks and field borders in dry soils: B–E, G–K, M–U, W–Z, (b), c–f, j–n, q–z, 125. 'Pastures and cornfields' (Pitt 1796). (Map p 232).

Scabiosa columbaria L.   Small Scabious
Limestone rocks and banks: common in the Manifold Valley and
Dovedale: H–J, N–P, 19. Dovedale (Gisborne 1792!).

Succisa pratensis Moench   Devilsbit Scabious
Damp heathy pastures and rushy meadows: A–v, y, 345. Dickenson,
1798.

COMPOSITAE

Bidens cernua L.   Nodding Bur-marigold
By the sides of ponds, lakes and canals: frequent, particularly in
the centre of the county: B, E, G, K–L, N–U, W–b, d–e, (f), g, i–m,
(n), r, t, w, 83. 'In a splashy rivulet at the bottom of Tittensor
Common' (Stokes in Withering 1787). (Map p 232).

B. tripartita L.   Trifid Bur-marigold
'Pits and watery places' (Dickenson 1798) and by the sides of
canals and reservoirs: frequent, particularly in the south: B, (F),
G, K–M, (N), Q–R, T–U, X–b, d–h, k–t, w–y, 81. It varies in
quantity according to the season and the state of the ground. In
1911 Ridge found it in profusion on the drying slopes of Deep
Hayes Reservoir. In June he had seen only an odd plant or two,
but in the late summer there were 'enormous quantities' on the
banks of the reservoir. (Map p 232).

B. frondosa L.   Beggar-ticks
A newcomer which is spreading along the canals in the south east:
k–p, t, x, 10. Canal cutting near Hamstead, SP/035939, W. H.
Hardaker (1954!); canal side at Rough Wood, Short Heath
(1964!); canal at Hopwas (1969).

Galinsoga parviflora Cav.   Gallant Soldier
Another newcomer. Allotments at Burton (Burges, Hardaker,
Lousley and Thomas in BEC 1936 Rep); near Bolehall, Tamworth
(1958); bean field at Wordsley (1959!); abundant in a garden in the
centre of Hanley (1961).

Senecio jacobaea L.   Common Ragwort
Sandy heaths, waysides and rough pastures, especially on disturbed
ground in base-rich soils: B–y, 545. Pitt, 1796.

S. aquaticus Hill   Marsh Ragwort
Marshy fields: B–s, v–y, 260. Pitt, 1796. (Map p 232).

S. eruciflious L. Hoary Ragwort
Hedgebanks, chiefly on the Keuper marl: local: (K–L), (N), P,
T–U, W–b, (d), m–w, 37. Blymhill (Dickenson 1798).

S. squalidus L. Oxford Ragwort
Slag heaps, pit mounds, walls and railway sidings in the industrial
areas: abundant in the Black Country: C, E–H, (K), L–N, S,
U–V, X–b, d–u, w–y, 250. 'Sir Roger Curtis and myself noticed this
in great quantity on the slag heaps near Walsall, Bentley Common,
Ocker Hill and Tipton, to which it had spread from the adjoining
railway. It was also seen near the Wren's Nest and on top of
Dudley Castle' (Druce in BEC 1923 Rep). Away from the towns
odd plants are sometimes found by the roadside. *S. squalidus* x
vulgaris is recorded for Burton (BEC 1927 Rep) and *S. squalidus*
x *viscosus* for Norton Canes (1948!). (Map p 233).

S. sylvaticus L. Heath Groundsel
'Dry heaths and sandy ditch banks' (Dickenson 1798): frequent:
B, E–G, K–L, (M), N, Q–U, W–g, k–r, t, w, 126.

S. viscosus L. Sticky Groundsel
Rubbish dumps, rubble, pit mounds, railway tracks: frequent:
B–C, F–H, K–P, R–V, X–b, d–h, k–u, y, 159. By the railway at
Milford and Brocton station (Reader 1920!). Has increased
considerably during the last fifty years. In 1844 Garner said it was
common, but he was doubtless mistaken. There are no certain
records for the nineteenth century.

S. vulgaris L. Groundsel
Arable and waste land everywhere: A–z, 778. Pitt, 1794. The rayed
form, var *hibernicus* Syme, is often seen but has not been separately
recorded.

S. fluviatilis Wallr. Broad-leaved Ragwort
On 7 September 1832 Babington wrote in his diary, 'At Thatchmore,
near a house, but my uncle says quite wild' (A.M.B. 1897); hedges
near Compton mill and at Bradwell (Pitt 1817); old garden at
Wolstanton (Garner 1878).

Doronicum pardalianches L. Leopard's-bane
Woods and shady roadsides: naturalised in a few places: B, (L),
(N), R–U, Z, j, v, 8. Dimsdale (Spark in Garner 1844).

Tussilago farfara L.   Coltsfoot
'Arable land in a clayey soil; a very noxious weed where it pre-
dominates and difficult to eradicate; frequent tillage encourages
its growth' (Dickenson 1798): also common on waste ground in
derelict town areas: A–z, 741. Pitt, 1794.

Petasites hybridus (L.) Gaertn., Mey. and Scherb.   Butterbur
Riversides, especially in wet flat places where there is a junction of
streams: common in the north and abundant in the Manifold
Valley, where it spreads across the river bed in a dry season:
A–P, R, T–b, d–f, h–k, n, r–t, w, 180. 'In Mothersea Brook near
Stone' (Withering 1787). The 'female' plant with tall stems of
plumed fruits is known for the following squares: A–C, E, G–J,
N–P, Y, a, e, j, r. (Map p 233).

P. albus (L.) Gaertn.   White Butterbur
Plentiful in Star Wood, Oakamoor, and near Alton, by the river,
between Holm Cottage and Lord's Bridge. It looks like a native
plant, but as it is not mentioned by Garner, who knew this district
well, we must suppose it to have been introduced. It was first
reported by one, Bostock, in 1889. The following year Miss Evans
of Seabridge Hall found it at Butterton, near Newcastle, where it
was rediscovered in 1948 in a dry ditch on the west side of the
nurseries.

P. fragrans (Vill.) C. Presl   Winter Heliotrope
Garden escape. Eccleshall, H. V. Thomspn (1952!); Dunstall (1956!).

Inula helenium L.   Elecampane
A garden plant recorded for Himley Wood (Bree in Purton 1817)
and Biddulph Castle (Garner 1844).

I. conyza DC.   Ploughman's-spikenard
Slag heaps, wood clearings, disused railway tracks, road and canal
sides and rocky hill slopes in dry base-rich soils: rare on the
Carboniferous limestone, frequent between Dudley and Kinver:
(H), J, r–s, w–x, 13. Dudley Castle and roadside to Kingswinford
(Wainwright in Shaw 1801). There are only two records for the
north. Ridge found it in the Manifold Valley in 1919 (Flora) and
Miss Hollick found one piece on Bunster Hill in 1966. (Map p 233).

Pulicaria dysenterica (L.) Bernh.   Common Fleabane
Marshy places, chiefly in the south: (F), G, K, (M–N), P, R–S,
U–X, (Y), Z–b, d–f, j–s, v–w, y, 58. Pitt, 1796. (Map p 233).

Filago vulgaris Lam. Common Cudweed
Sandy fields and gravel pits: rare: W, Y, (Z), (b), (f–g), r, w, 4.
Pitt, 1796. Old records for Burton, Lichfield, Rugeley, Trysull and
Kinver. Recent records for Baswich, sandy heath (1944!);
Wombourn, sand pit (1954!); Forton, growing with *Potentilla
argentea* on the hillside above Windswell Pool (1954!); Gibbet
Lane, Kinver (1960!).

F. minima (Sm.) Pers. Small Cudweed
Dry open sandy places: frequent in the south west, but rare everywhere else. Weston-under-Lizard (Dickenson 1798); Tittensor
(Garner 1844); Highgate Common (Fraser 1865!); Whittington
near Kinver in 1877 (Mathews 1884); pool near Himley Wood
(Bagnall 1901); Milford, gravel pit (Thornton 1924!); gravel pit
near Great Haywood (1925); Trysull, sand quarry (1946!); sand pit
near Smestow Gate (1952).

Gnaphalium sylvaticum L. Heath Cudweed
Woodland rides in dry acid soils: frequent formerly, if not today,
in the Churnet Valley and near Kinver, but rare elsewhere: (H),
(K), L, (N–P), R, (T), (d), (f), (n), w, 5. Blymhill (Dickenson 1798).
Garner (1844) said it was common in gravelly places. Recent
records for Enville (1946!); Bishop's Wood (1951!); Kinver
(1954!); Trentham Park.

G. uliginosum L. Marsh Cudweed
Damp hollows in ploughed fields, tracks through woods and the
muddy edges of ponds: B–C, F–G, K–P, R–u, w, y–z, 264.
Dickenson, 1798.

Anaphalis margaritacea (L.) Benth. Pearly Everlasting
Near Lichfield (Withering 1801); hedges at Halfhead (Garner
1882); waste ground by canal at Lower Penn (Fraser 1884!);
Morridge just beyond Leek, a clump of 300 plants growing with
Juncus effusus well away from the road (Ridge 1915!).

Antennaria dioica (L.) Gaertn. Mountain Everlasting
'Moorlands, on high hills' (Bourne in Dickenson 1798); 'Limestone
hills, Wetton Valley' (Garner 1844).

Solidago virgaurea L. Golden-rod
Shady banks and rocky woods on sandstone and gritstone: frequent
and locally common in the north, as in the Churnet Valley and at

Three Shires Head: B–C, (F), G–H, K–N, R, (S), T–U, Y, e, (n), (r), (t), (x), z, 47. 'Plentifully throughout the park at Willowbridge' (Waring 1770). (Map p 234).

Aster tripolium L.   Sea Aster

First recorded by Plot (1686) for a salt marsh near Ingestre. Stokes (Withering 1787) said it grew in a meadow between the Trent and the canal and Wolseley (Dickenson 1798) in a salt marsh near Tixall. Plot (p 202), says, 'It grows here in a ground called the Marsh, near the place where the brine of its self breaks out above ground, frets away the grass, and makes a plash of salt water.' On another page (p 97) he writes, 'The subterraneous brine is so strong, that the cattle standing in it in summertime and throwing it on their backs with their tails, the sun so candies it upon them that they appear as if covered with a hoar frost.' There is one other old and unconfirmed record: 'At the entrance of Hollow meadow, at Branston' (Riley in Shaw 1801).

Erigeron acer L.   Blue Fleabane

Recorded for walls with lime-rich mortar, limestone rubble, gravel pits and waste sandy ground: rare: (G), (K), a, (b), (d), g, r–s, (t), w, 5. 'Stretton bridge on the Watling Street way; it grows out of the mortar betwixt the joints of the stones on both sides of the bridge' (Dickenson 1798). Other old records for Dudley Castle, Tutbury Castle, Kinver, Great Barr, Whitmore and Cheddleton Heath. Recent records for disused railway track in Himley Wood (1946), gravel pit along Catholme Lane, Wychnor (1956!), and the explosion pit at Fauld, Hanbury (1970).

Conyza canadensis (L.) Cronq.   Canadian Fleabane

Gravel pits and sandy waste ground: frequent and increasing, especially in the south west: G, (L), S, U, Y–Z, b, f, j, m, p, r, (w), 16. Ashwood (Wainwright in Shaw 1801).

Bellis perennis L.   Daisy

Pastures and lawns: A–z, 775. 'Gathered by me in 1783' (Gisborne!).

Eupatorium cannabinum L.   Hemp-agrimony

Recorded for the banks of canals and rivers, but chiefly a plant of ditches: uncommon in the north, frequent in the south: E, (F), G, J–L, (N), P–S, U, W–a, (b), c–g, j–u, w, 53. 'In moist ditches' (Pitt 1794). (Map p 234).

Anthemis tinctoria L.   Yellow Chamomile
Casual. Waste ground near Walsall (Curtis in BEC 1927 Rep); waste ground, Silverdale (1961).

A. cotula L.   Stinking Chamomile
Arable and waste land, particularly in clay soils: frequent and locally common: B–D, G–L, N–Q, (S), T–U, W–Z, (b), c–g, j–m, r–s, v–w, z, 70. Pitt, 1794. 'Frequently so abundant in cornfields as greatly to injure the crop' (Dickenson 1798). (Map p 234).

A. arvensis L.   Corn Chamomile
Arable fields: rare. It is impossible to say when this species was first recorded for Staffordshire, because the early botanists seem to have misunderstood it. But the record for Kinver, 1862 (Mathews 1884), is probably reliable. Fairoak near Eccleshall (1954!); Balterley (1961!).

Chamaemelum nobile (L.) All.   Lawn Chamomile
Recorded for commons, roadsides and lawns in the past, but there are no recent records. 'On old trodden turf' (Pitt 1794). 'In the road between Cannock and Rugeley, a little to the south west of the Hedgford finger-post' (Pitt 1796). Blymhill lawn in great abundance (Dickenson 1798); Lichfield (Jackson 1837); Chorley, side of brook near public house (Power MS); Longdon Green about 1840 (Reader 1922); once near Sudbury and once near Stafford (Ridge in Flora).

Achillea millefolium L.   Yarrow
Pastures and roadsides: A–z, 793. Pitt, 1794.

A. ptarmica L.   Sneezewort
Marshy places in meadows, by roadsides and on the moors: B–D, F–P, R–b, d–g, j–t, w–z, 235. Pitt, 1796. (Map p 234).

Tripleurospermum maritimum (L.) Koch subsp *inodorum* (L.)
Hyland. ex Vaarama   Scentless Mayweed
Cultivated fields, building sites, pit mounds and rubbish dumps: B–z, 681. Garner, 1844.

Matricaria recutita L.   Scented Mayweed
Arable fields and waste ground: B–C, E–P, R–u, w–y, 323. Dickenson, 1798. (Map p 235).

M. matricarioides (Less.) Porter   Pineapple-weed
Ground subject to trampling, such as roadsides, farmyards, tracks

and gateways: A–z, 787. This now abundant weed was first recorded for Britain in 1871 (Druce 1932). A plant collected by Nowers at Burton in 1888 is probably the first record for Staffordshire.

Chrysanthemum segetum L. Corn Marigold
Arable fields in light sandy soils: local, but sometimes abundant in neglected fields: particularly common about Rugeley: C, (F), G, (J), (L–N), R, (S), W, (Y), Z, b, f–g, j, m, p–s, w, 35. Pitt, 1794. 'In cornfields, a specious flower, but pernicious weed' (Dickenson 1798). (Map p 235)

Leucanthemum vulgare Lam. Oxeye Daisy
Hayfields, pastures, railway embankments and newly made roadside banks and verges, preferring base-rich soils: A–D, F–P, R–z, 388. Dickenson, 1798.

Tanacetum parthenium (L.) Schultz Bip. Feverfew
Hedges and waste ground near dwellings, but particularly common on walls: probably a naturalised garden escape: widespread: B–C, F–P, R–U, W, Y–b, d–s, v–w, y–z, 88. Pitt, 1794.

T. vulgare L. Tansy
Frequent on waste ground by the roadside, but most at home on river banks: common along the Trent and Dove: F–H, K–M, R–z, 141. Pitt, 1794. (Map p 235).

Artemisia vulgaris L. Mugwort
Waste ground, especially in town areas, and by roadsides but, as Purchas (1885) observed, almost completely absent from the Carboniferous limestone: C, E–H, K–g, i–z, 504. Pitt, 1794.

A. absinthium L. Wormwood
'Roadsides and about villages in a light soil' (Dickenson 1798): F–H, L–M, (R), S–V, Y–b, d–y, 240. This plant has increased greatly during recent years. Garner in 1844 and Ridge eighty years later considered it rare. It is now common on waste ground in the industrial districts and abundant in the Black Country. (Map p 235).

Carlina vulgaris L. Carline Thistle
Hillsides and old quarries on the limestone and banks in disused collieries on Cannock Chase: local: J, (L), N, f, m, (s), t, (w), 8. Dickenson, 1798.

Arctium lappa L. Greater Burdock
Rubbish dumps and waste places, sometimes at the edge of a wood

and often near water: rare. Wren's Nest, Sedgley (Bagnall 1890!);
Penkridge, roadside near Whiston (1944!); Marchington, rubbish
dump between the village and the railway (1948!); West Bromwich
(1948); Kinver, canal side (1954); Upper Arley, riverside (1954);
Barton, rubbish tip along Bar Lane (1962).

A. minus Bernh. sensu lato   Lesser Burdock
Woods, thickets, roadsides and waste ground: B–C, E–y, 456.
Pitt, 1794. The following account of the segregates rests entirely on
specimens determined for me by Dr F. H. Perring in 1964.

A. nemorosum Lejeune
Cheadle (Masefield 1883!); Dimminsdale (Masefield 1883!);
Ellastone (1942!); Newcastle, rubbish dump off Clayton Lane
(1945!); Star Wood (1963!); Rocester, at bridge over Dove (1963!);
Wootton, milestone, SK/1044 (1963!).

A. pubens Bab.
Grindon, in the Manifold Valley (1941!); Dimminsdale (1942!);
Wetton (1944!); Hanbury, roadside by Greaves Wood (1945!);
Waterhouses, in the Manifold Valley (1945!); Marchington, rubbish
tip (1948!); Penkridge, roadside near Whiston (1963!); Swynnerton,
the Stretters (1963!); King's Bromley, canal bridge (1963!);
Norbury, Mill Haft (1963!); Sedgley, Baggeridge Wood (1963!).

A. minus Bernh. sensu stricto
Baswich (1944!); Hanbury, roadside by Greaves Wood (1945!);
Swinfen and Packington (1945!); Kinver (1954!); Abbots Bromley
(1963!); Highgate Common (1963!); Gnosall, wooded canal
cutting (1963!); Ranton, Stubbs Wood (1963!); Seighford, waste
ground (1963!); Brewood, near the village and roadside north of
Langley Lawn (1963!); Patshull, Jubilee Plantation (1963!);
Blymhill, Brockhurst Coppice (1963!).

Carduus tenuiflorus Curt.   Slender Thistle
Casual. Burton (Curtis and Druce in BEC 1928 Rep).

C. nutans L.   Musk Thistle
Roadsides, fallow fields and quarries in dry base-rich soils: common
on the Carboniferous limestone and in the south west: C, G–J,
M–Q, S, W–Y, (a), b–n, r–t, w–x, 98. 'Blymhill, so abundant in
clover fields in light land as to be a troublesome weed' (Dickenson
1798). (Map p 236).

C. acanthoides L. Welted Thistle

Riversides: the records provide us with a map of the Trent, the Manifold and the lower reaches of the Dove: H–J, (N), P, (S), T–V, Y–b, d, f–h, k, n, r–t, v, (w), 53. Near Chartley Castle and near Burton (Dickenson 1798). (Map p 236).

Cirsium eriophorum (L.) Scop. Woolly Thistle

Rare on the Carboniferous, but frequently recorded for the Silurian limestone. Near Burlington in the parish of Sheriff Hales, ruins of Dudley Castle and plentifully by the roadside between Wednesbury and Bilston (Dickenson 1798); Wren's Nest, Sedgley limeworks and Barrow Hill (Garner 1844); Hayhead (Fraser 1864); among bushes at the foot of Bunster Hill (1947!); Upper Arley, several fine plants in the lane to Hextons Farm (1954!).

C. vulgare (Savi) Ten. Spear Thistle

Fields, roadsides and waste ground: A–z, 793. Pitt, 1794.

C. palustre (L.) Scop. Marsh Thistle

Wet places in many habitats: A–z, 682. Pitt, 1794.

C. arvense (L.) Scop. Creeping Thistle

Meadows, arable fields and waste ground: A–z, 796. Pitt, 1794.

C. acaule Scop. Dwarf or Stemless Thistle

Hillside behind Wetton Mill, J. B. Pendlebury (1971).

C. heterophyllum (L.) Hill Melancholy Thistle

In the dales and by roadsides on the Carboniferous limestone: C–D, H–J, (P), 14. Garner, 1844. Still plentiful by the old railway track near Thor's Cave, where Ridge (1913) said it was 'fairly abundant.'

C. dissectum (L.) Hill Meadow Thistle

'Blymhill, in marshy meadows of a bad quality' (Dickenson 1798): local: P, (R), W–X, (a), d, m, (n), t, 10. Stanton, rough pasture where peat overlies limestone (1934!); in the fritillary field at Wheaton Aston (1949!); water meadows on the west side of Aqualate Mere (1953!); Norton Canes, with *Molinia* in a moorish swampy field, SK/045053 (1962); Allimore Green Common (1967); Doley Common in wet peat (1968).

Silybum marianum (L.) Gaertn. Milk Thistle

Casual. Alton, near Stone, Burton (Garner 1844); Walton and

Shobnall (Brown 1863); railway cutting near Rugeley in 1897 (Bagnall 1901); Oakamoor, garden (Berrisford 1903!); Burton (BEC 1928 Rep); garden at Betley, springing up spontaneously to the astonishment of the gardener (1964).

Onopordum acanthium L. Cotton Thistle
Recorded as a casual or garden outcast for Aqualate Park (Garner 1844); Burton (Nowers 1907!); Wigginton Lodge (1964!).

Centaurea scabiosa L. Greater Knapweed
Railway banks and grassy roadsides in dry calcareous soils: rare in the north, frequent in the extreme south: J, (K), (N), (R), S, (Y), (b), j–n, r–t, (v), w, 24. Ilam (Pitt 1794). 'Borders of cornfields; about Tutbury, frequent' (Dickenson 1798). (Map p 236).

C. cyanus L. Cornflower
Cornfields (Dickenson 1798): formerly frequent, now rare: (G), H, (K–N), (R–S), T, (W), (Y), Z–a, (b), (g), (t), 4.

C. nigra L. sensu lato Common Knapweed
Grassland: A–z, 769. Stokes in Withering 1787.

C. nemoralis Jord Slender Knapweed
This taxon is less common than *C. nigra* sensu stricto, but it has not been separately recorded. Cannock Chase (Druce in BEC 1922 and 1923 Reps).

C. calcitrapa L. Red Star-thistle
Casual. Hill above Kingston Pool (Clifford 1817); Burton (Nowers 1896! and Curtis 1930).

C. solstitialis L. Yellow Star-thistle
Casual. 'In waste ground' (Garner 1844); Oakamoor, railway embankment (Berrisford and Walker 1906); Burton (BEC 1927 Rep); Horninglow near Burton, a few plants in an arable field (1968!); garden at Trentham (1970).

Serratula tinctoria L. Saw-wort
Rough damp pastures and bushy hillsides, often growing with gorse in loamy soil over limestone and on the Keuper marl: local, with a scattered distribution: (G), H, P, (S–T), (Y), Z–b, d, w, 9. 'Stone Field by the road leading from thence to the Out Lanes' (Forster 1796). 'Woods and marshy meadows' (Dickenson 1798).

Cichorium intybus L. Chicory
Recorded for cornfields, clover fields, canal banks, sand pits and

waste ground, but usually only a few plants are seen: probably always introduced accidentally or deliberately, though long known in the county: B–C, G, L, M, (N), P, S, (T), U, W–X, Z–b, d–e, (g), m, p, r, (w), x, 26. Tamworth Castle (Withering 1796). 'Roadsides and borders of cornfields, rare' (Dickenson 1798). Garner (1879) called it a straggler.

Lapsana communis L.   Nipplewort
Hedgerows, walls and waste ground: A–z, 683. 'Gardens and cultivated land, frequent' (Dickenson 1798).

Hypochoeris radicata L.   Catsear
'In meadows and pastures, common' (Dickenson 1798): A–z, 767. 'Thatchmore gravel pit 1791' (Gisborne MS).

H. glabra L.   Smooth Catsear
'Gravelly places' (Garner 1844): rare. Gravelly field between Kinver and Kinver Edge (Fraser 1877!); foot of Bunster Hill (Nowers 1909!); Brewood, a few plants in a disused sand quarry, SJ/844075 (1961!).

Leontodon autumnalis L.   Autumn Hawkbit
'Meadows and pastures' (Dickenson 1798): A–z, 688.

L. hispidus L.   Rough Hawkbit
Pastures and roadsides throughout the county in dry base-rich soils: abundant on the Carboniferous limestone: B–D, F–b, d–w, y–z, 254. Near Yoxall Lodge (Gisborne 1791!). (Map p 236).

L. taraxacoides (Vill.) Mérat   Lesser Hawkbit
Gravel pits, heathy banks, tracks through woods, in dry sandy soils, often with *Centaurium erythraea*: frequent: (A), G–H, L, N, S, (T), U, Y–a, e, (f), g, k, n, r–s, (t), y, 21. Bagnall, Barlaston Common, Wolstanton Marsh (Garner 1844). Purchas (1885) remarks on its rarity on the Carboniferous limestone. He found it on one of the limestone slopes on the Staffordshire side of Dovedale only after a special search.

Picris echioides L.   Bristly Oxtongue
Casual. Swynnerton war factory, once seen on waste ground, Miss J. Matthews (1943); Newcastle (1970).

P. hieracioides L.   Hawkweed Oxtongue
Limestone slopes and quarries, gravel pits, roadsides and railway

banks: local and almost confined to the eastern border of the county: H, L, (N), P, b, g–h, s, 8. Tutbury, Ecton Hill, Dovedale (Garner 1844). Recent records: Harpfields, Stoke (1940); Stanton (1941!); limestone rocks near Thor's Cave (1946!); abundantly on the steep embankment at Elford station (1946!); Sedgley, limestone quarry (1950!); Tutbury churchyard (1956!); Wychnor, gravel pit along Catholme Lane (1956!); Edingale, roadside half a mile east of the village (1957!).

Tragopogon pratensis L.   Goatsbeard
'Pastures in a clayey soil' (Dickenson 1798), railway embankments, canal sides, lime quarries and waste ground: B, E–Q, S–z, 300. The common plant is subsp *minor* (Mill.) Wahlenb, with florets conspicuously shorter than the involucral bracts, but plants with florets and bracts of equal length have been seen at Kingswinford (1946) and Calton (1947!).

T. porrifolius L.   Salsify
Garden escape. Unconfirmed records for Rolleston (Mosley in Garner 1844); Weston-under-Lizard (Garner 1844); Tixall (Clifford 1817); Tamworth (Bagnall 1901).

Lactuca serriola L.   Prickly Lettuce
Casual. Burton (Harlond 1941!); Blymhill (1961!).

L. virosa L.   Great Lettuce
Unconfirmed and perhaps doubtful records for Chartley Castle (Dickenson 1798); Hamstall Ridware (Shaw 1801); Tatenhill to Branston (Shaw 1801); Tutbury (Garner 1844); near Walton railway station (Brown 1863).

Mycelis muralis (L.) Dumort.   Wall Lettuce
Old walls, rocks and shady hedgebanks, especially on the Carboniferous limestone: B, F–b, d, f–g, j, n, r–w, z, 101. Near Yoxall Lodge (Gisborne 1791!). (Map p 237).

Sonchus arvensis L.   Perennial Sow-thistle
'Cornfields and ditch banks' (Pitt 1796) and 'often by canal sides' (Ridge in Flora): B–C, F–H, K–b, d–y, 326.

S. oleraceus L.   Smooth Sow-thistle
'Cultivated land, gardens and dung hills' (Dickenson 1798) and by roadsides: B–C, E–z, 526. Pitt, 1796.

S. asper (L.) Hill   Prickly Sow-thistle
Arable and waste land: A–z, 450. Trysull (Fraser 1863!).

Cicerbita macrophylla (Willd.) Wallr.   Blue Sow-thistle
Garden escape or outcast. Back of Ecton, Miss Hollick (1947);
Uttoxeter (1949!); Longdon Green (1951); Great Haywood
(1964); Newcastle (1967); Quixhill (1971).

Hieracium L.   Hawkweed
The following account of the Staffordshire hawkweeds rests on
specimens determined by P. D. Sell and Dr C. West. Most of them
are in my own herbarium, but some are at Cambridge and in other
collections. The best place to look for hawkweeds in Staffordshire
is the Manifold Valley, where most of the species in the following
list are native in the woods or on the rocks. Elsewhere hawkweeds
grow by roadsides, on railway embankments and especially and
sometimes in profusion on banks and waste ground about coal tips
and iron works.

H. subplanifolium Pugsl.
Rocks near Thor's Cave (1946!).

H. britannicum F.J.Hanb.
Near Alstonfield, W. H. Purchas (1894!); Sugarloaf (1948!);
Wetton Mill, T. G. Tutin (1949!); Stanton, north of Brown Edge,
E. Milne-Redhead (1947!); Thor's Cave (1951!); Ravens Tor near
Alstonfield, C. West (1951!); rocks by the Manifold below Thor's
Cave, SK/098550, P. D. Sell (1954!); Hall Dale (1964!). This
hawkweed is of special interest because it is probably restricted to
the limestone dales of Derbyshire and Staffordshire, but there it is
frequent and its large flowers are particularly attractive.

H. cymbifolium Purchas
'This hawkweed has been known to me for at least the last fifteen
years. I first noticed it on rocky (limestone) banks in the valley of
the Manifold, near Wetton' (Purchas 1899). Wetton Mill, W. R.
Linton (1896!); Alstonfield, A. Ley (1906!); Thor's Cave (1944!);
roadside near Wetton Mill, SK/102542, Sell (1954!).

H. exotericum Jord. ex Bor. sensu lato
Manifold Valley, north of Wetton Mill, V. S. Summerhayes (1947!);
Maer, wall by the Hall (1951!); below Thor's Cave, SK/095549,
Sell (1954!); Butterton, between the tunnel and Wetton Mill (1964!).

H. cinderella (A.Ley) A.Ley
Seckley Wood (1954!).

H. sublepistoides (Zahn) Druce
Hamps Valley (1932!); Maer, roadside between the War Memorial
and the Hall (1950!); Hoar Cross near Yoxall (1956!).

H. subprasinifolium Pugsl.
Ravens Tor, J. E. Woodhead (1947!); Hall Dale (1964!).

H. rubiginosum F.J.Hanb. var *peccense* W.R.Linton
Hall Dale (1947!); Wolfscote Dale (1953!); rocks by Wetton Mill,
SK/096562, Sell (1954!).

H. decolor (W.R.Linton) A.Ley
Manifold Valley, by the old railway, SK/097555, Sell (1953!).

H. caledonicum F.J.Hanb.
Wetton Mill (1964!); Hall Dale (1964!).

H. vulgatum Fr.
Wall Grange (1938!); Knutton (1944!); Fawfieldhead (1944!);
Brocton (1945!); Quarnford (1947!); Penkridge, railway embankment east of Mansty Wood (1954!); Dilhorne, railway (1958!);
Trentham gravel pit (1959!); site of war factory near Swynnerton
(1961!); Stoke, Sideway, spoil heaps and railway sides (1964!);
and in many places on the limestone.

H. maculatum Sm.
Codsall (Fraser 1882!).

H. diaphanum Fr. (including *H. anglorum* (A. Ley) Pugsl.)
Railway sidings, spoil heaps, sandy roadsides and hillsides on the
limestone: F–J, L, S, X–Y, b, f, k–n, r, 19.

H. strumosum (W.R.Linton) A.Ley
Roadsides, hillsides, rocks and walls: C, F–J, L, N, Y–Z, f, s, w, 22.
Cheadle (Masefield 1883!).

H. subcrocatum (E.F.Linton) Roffey
Apart from one record a plant of the hills. Near Stafford, Ley
(1902!); Reaps Moor, Ley (1906!); Quarnford, near Gib Torr
(1947!); Manifold Valley, E. M. Rutter (1950!); Fawfieldhead
(1952!); Apes Tor (1964!).

H. umbellatum L.

Needwood Forest, probably C. C. Babington (1839!); Alstonfield, corner of plantation on Archford Moor, probably W. H. Purchas (1878!); Bishop's Wood (1946!); Himley, railway embankment (1946!); railway embankment between Lichfield and Wall (1948!); Mavesyn Ridware, near Bentley Farm (1950!); Wombourn, sand quarry (1954!); Penkridge, roadside, SJ/915154 (1957!); railway near Onneley (1958!); Maer (1963!); Rough Wood, Short Heath (1964!).

H. perpropinquum (Zahn) Druce

Frequent in the north in dry heathy places: C, G–H, K–L, N, R, U, X, a, c, k, r, 21.

H. salticola (Sudre) Sell and West

Frequent in the south. Burton, railway sidings, J. E. Lousley (1936!); Burnt Wood (1947!); Manifold Valley, E. M. Rutter (1950!); Abbots Bromley, SK/090290 (1956!); Walsall Wood, canal side (1957!); Enville Common (1963!); Wombourn (1964!); Essington, Mill Farm (1964!); Brewood, Paradise to Brinsford (1964!); Wychnor, Catholme gravel pit (1964!); Wigginton (1965!).

H. vagum Jord.

By streams in woods and by roadsides: frequent: B–C, G, J–R, b, e–g, k, p, 22.

H. pilosella L.   Mouse-ear Hawkweed

Dry grassy roadside banks, wall tops and hillsides: A–z, 424. Near Yoxall Lodge (Gisborne 1792!). Of seventeen Staffordshire specimens in my herbarium one from Wetton (1947!), which had long and short glandular hairs on the phyllaries, was a distinct but undescribed variety and eight were named var *pilosella*. The rest were determined as follows: var *concinnatum* Hanb., Wetton Mill (1947!); Blore, SK/1249 (1958!); Bunster Hill (1965!); Weaver Hills, SK/0946 (1965!): var *nigrescens* Fries, Ecton, SK/0958 (1955!): var *tricholepium* (NP.) Pugsl., Thor's Cave slope (1964!); Swynnerton, SJ/8432 (1964!); Bunster Hill (1965!).

H. peleteranum Mérat var *tenuiscapum* Pugsl.

Rocks by Wetton Mill, Sell (1954!).

H. brunneocroceum Pugsl.   Orange Hawkweed

Naturalised on railway embankments and in churchyards and

neglected corners of parks and gardens: frequent: B–C, G–J, L–P, R–U, W–b, g, j, m–n, y, 36. Railway embankment below Burton (Brown 1863).

Crepis vesicaria L. subsp *taraxacifolia* (Thuill.) Thell.
Beaked Hawksbeard
Uncertain. Ridge (1918) claimed to have seen it in many places, but he may have been mistaken. Druce (BEC 1922 Rep) found it at Patshull but in Shropshire.

C. capillaris (L.) Wallr. Smooth Hawksbeard
Roadsides, arable and waste land: B–z, 474. Dickenson, 1798. Var *glandulosa* Druce is common.

C. paludosa (L.) Moench Marsh Hawksbeard
Marshy places by moorland streams, often most luxuriant in the cloughs where there are sheltering trees or protecting rocks: south of the hills it grows in shady places by streams and pools but is rare: B–C, F–H, (L), N–P, R, (S), T, f, 32. 'On the banks of a rivulet near Biddulph' (Pitt 1817) and in the marshy wood on the east side of Knypersley Pool (1938!), which may well be the same place. (Map p 237).

Taraxacum officinale Weber Dandelion
An abundant and beautiful flower in grassy places in the spring: A–z, 793. Pitt, 1794. Dr A. J. Richards has examined the specimens in my herbarium and we can now claim the following twenty-nine microspecies for the county:
Section *Vulgaria*: *T. adsimile* Dt., *T. crispifolium* Lindb. f., *T. dahlstedtii* Lindb. f., *T. dilatatum* Lindb. f., *T. ekmanii* Dt., *T. fasciatum* Dt., *T. hamatum* Raunk., *T. hamatiforme* Dt., *T. lingulatum* Markl., *T. longisquameum* Lindb. f., *T. marklundii* Palmgr., *T. pallescens* Dt., *T. polyodon* Dt., *T. laciniosum* Dt., *T. sublaciniosum* Dt., *T. tenebricans* Lindb. f.
Section *Spectabilia*: *T. adamii* Claire, *T. euryphyllum* (Dt.) M.P.Chr., *T. faeroense* (Dt.) Dt., *T. maculigerum* Lindb. f., *T. nordstedtii* Dt., *T. spectabile* Dt., *T. unguilobum* Dt. Dandelions belonging to this section, which in the aggregate are sometimes given the name red-veined dandelion, occur frequently in wet rushy meadows, particularly among the hills: B–D, F–P, R–S, W–X, Z–a, (b), d–g, n, (r), (z), 65. Early records under the name *T. palustre* probably belong here.

Section *Erythrosperma: T. brachyglossum* (Dt.) Dt., *T. lacisto-
phyllum* (Dt.) Raunk., *T. oxoniense* Dt., *T. pseudolacistophyllum*
V.S., *T. rubicundum* (Dt.) Dt., *T. fulvum* Raunk. The *Erythrosperma*,
collectively called lesser dandelion, are plants of dry limestone
pastures and sandy heaths, where the soil is shallow and the turf
short: H–J, N–P, (U), Z, (b), (f–g), (j), (n), (r–t), 19. The earliest
specimen I have seen came from Dovedale (Masefield 1883!) and
has now been determined *T. rubicundum*.

ANGIOSPERMAE: MONOCOTYLEDONES
ALISMATACEAE

Baldellia ranunculoides (L.) Parl.   Lesser Water-plantain
Whitemere Bog (Gisborne MS undated); 'Plentifully in a lake of
water called Rome Pool near the Gorse, on Knightley Common'
(Forster 1796); 'Pitmoor Pool, Weston-under-Lizard; in a pit in
the glebe Motty-meadow, Blymhill' (Dickenson 1798); marl pits
at Fradley (Power 1819!); Aqualate and near Burton (Garner
1844); Cannock Chase (Brown 1863); Allimore Green Common,
sparingly in a runnel (1967!).

Luronium natans (L.) Raf.   Floating Water-plantain
Canals and reservoirs: local. Near Kidsgrove, Ridge in 1914 (1917);
Cannock Chase Reservoir (Clark 1917!), in 1918 abundant and
flowering (Reader 1923); Norton Bog, plentiful in shallow water
(BEC 1925 Rep); canal at Whittington (BEC 1943 Rep); abundant
in the canal at Cheddleton, south of the station (1956) and between
the village and Wall Grange (1969!); canal at Tamhorn Park
Farm near Tamworth (1968!).

Alisma plantago-aquatica L.   Water-plantain
Ponds and canals, in mud and shallow water: B, E–H, K–P, R–u,
w–y, 321. Pitt, 1794. Plants with narrow leaves have sometimes
been recorded as *A. lanceolatum* With., but the occurrence of the
true species has not yet been established.

Sagittaria sagittifolia L.   Arrowhead
In ponds and ditches and near the banks of broad placid rivers, but
chiefly in canals: B, G, L–M, W, Y–Z, b, e–h, k, n–p, s, w–x, 57.
'In Lush Pool' (Gisborne 1791!). (Map p 237).

BUTOMACEAE

Butomus umbellatus L.   Flowering-rush
Ponds, rivers and canals, usually only a few plants in any one

place and they often out of reach in the deep water: frequent: G, L, P, (S), T, W, (X), Y, (Z), a, d–h, k–n, (r), s, u, w, 38. Stafford and Tamworth (Withering 7187). (Map p 237)

HYDROCHARITACEAE

Hydrocharis morsus-ranae L.   Frogbit
Unconfirmed records for Aqualate and Balterley (Garner 1844). Introduced at Madeley by T. W. Daltry (Fraser 1873! and H. W. Daltry 1911! and 1917).

Stratiotes aloides L.   Water-soldier
This was introduced into a pond at Madeley by T. W. Daltry and persisted there for many years (1917). Fraser gathered specimens from 'Field near Madeley' (Bagnall 1870! and Fraser 1873!) and H. W. Daltry from 'Pit at foot of Bar Hill' (1911!).

Elodea canadensis Michx.   Canadian Waterweed
Rivers, lakes, canals: widespread, but less common than it was seventy years ago: B, (F), G, J, L–N, S–T, W–Y, a–b, d, f–g, k–n, q, s–u, x–y, 52. 'It was in the summer of 1849 that this troublesome waterweed was first discovered in Derbyshire and Staffordshire, where it formed, as it were, small green meadows on the water, both in the Trent and the adjoining canals' (Pratt 1899). 'The lake (Trentham Lake) is now occupied by a very inveterate enemy. . . . It first made its appearance here about three years ago, and in that short time has spread entirely over the lake and the greater part of the river' (Garner 1857). 'Canals and rivers, very abundant' (Brown 1863).

JUNCAGINACEAE

Triglochin palustris L.   Marsh Arrow-grass
Marshy fields, sometimes plentiful but usually in small quantity and, when it grows in tall grass, easily overlooked: widespread: B–D, F–U, W–g, j–t, y, 112. Tamworth (Withering 1787). (Map p 238).

T. maritima L.   Sea Arrow-grass
'Salt marsh near Ingestre' (Stokes in Withering 1787). 'In a salt marsh near Tixall' (Wolseley in Dickenson 1798). 'Site of old salt marsh, Shirleywich' (Reader 1923!), plants two feet tall (Reader 1924), SJ/984256 (1947!). Recorded also for Branston meadows (Brown 1863), but Nowers could not find it there in 1889 (Burton Flora 1901).

POTAMOGETONACEAE

Mr J. E. Dandy has generously extracted the Staffordshire records from the card index of specimens authenticated by him and Sir George Taylor. These are the only records, apart from those of *P. natans*, included in the following account. Most of the specimens, except my own, are in the herbarium of the British Museum.

Potamogeton natans L.   Broad-leaved Pondweed
Canals and rivers, but most often seen on ponds and lakes: B–C, E–P, R–p, s–u, w–y, 225. 'In a pit in the Foxholes' (Gisborne 1792!).

P. polygonifolius Pourr.   Bog Pondweed
Wet hollows on the moors and in drains cut through the peat: probably frequent among the hills. Cannock Chase (Brown 1861!); Pound Green (Duncan 1897!); Moss Carr (1966!).

P. lucens L.   Shining Pondweed
River Trent at Stapenhill (Hind 1849!); Burton (Brown 1861!); river Tame at Elford (Hiern 1865!); Gailey Pools (Daltry 1924!); canal near Tamhorn Park Farm (Arnold 1968!).

P. x *zizii* Koch ex Roth (*P. gramineus* x *lucens*)
Stafford (Douglas 1850!); river Trent (Brown 1861!); Cannock Chase (Druce 1925!).

P. alpinus Balb.   Reddish Pondweed
Ponds and gently flowing streams. Madeley (Daltry 1840!); Stone (Stratton 1866!); Rudyard Reservoir (Searle 1884!); canal feeder, Brown Edge (Painter 1888!); canal feeder, Norton-in-the-Moors (Painter 1888! and Edees 1949!); Biddulph (Painter 1890!); The Serpentine, Knypersley (Thompson 1890!); pond near Harborne (Jordan undated!); Squattlesea, Idlerocks, Stone (Wedgwood 1905!); canal at Wall Grange (1949!); canal feeder from Rudyard Lake (1949!); pool near Great Bridgeford (1951!).

P. praelongus Wulf.   Long-stalked Pondweed
River Sow, Stafford (Douglas 1850! and Kirk 1858!); canal, Shobnall (Hanbury 1861!); river Trent, Burton (Brown 1861!); canal, Burton (Gibbs 1894!); canal, Lichfield (Burges 1939!).

P. perfoliatus L.   Perfoliate Pondweed
A common species of canals in the south of the county: W, Y, g, m–p, s–t, w–x, 14. Birchills Canal near Walsall (Lowe 1834!).

P. friesii Rupr. Flat-stalked Pondweed
Canal at Armitage (Reader 1899!); Lichfield (Druce 1920!);
Stourbridge Canal, Pensnett to Wordsley, Kingswinford, Brierley
Hill (Cadbury 1949!); Forton (1949!); small pool against Harborne
Reservoir (Goodman 1959!); canal, Weston-upon-Trent (Howitt
1961!); canal, Fazeley (Arnold 1968!).

P. pusillus L. Lesser Pondweed
Canal at Pendeford (Fraser 1865!); pools near Parkfield iron works,
Rough Hills, Wolverhampton (Fraser 1882!); marl pits, Fradley,
Alrewas (Power undated!); Lichfield (Druce 1920!); river Trent,
Burton (Taylor 1937!); canal, Whittington (Arnold 1968!); canal
near Tamhorn Park Farm (Arnold 1968!).

P. obtusifolius Mert. and Koch Blunt-leaved Pondweed
'Lush Pool' (Babington 1832!); Abbots Bromley (1956!).

P. berchtoldii Fieb. Small Pondweed
Pool near Walsall (Lowe 1834!); The Serpentine, Knypersley
(Thompson 1890!); Madeley (Daltry 1911!).

P. trichoides Cham. and Schlecht. Hair-like Pondweed
Turner's Pool, Kettlebrook near Tamworth, SK/214029 (Arnold
1968!).

P. compressus L. Grass-wrack Pondweed
Long Wood Canal near Walsall (Lowe 1834!); Stafford (Douglas
1850!); canal at Fazeley (Oakeshott 1853!); canal, Shobnall
(Hanbury 1861!); river Trent, Burton (Harris 1880!); Lichfield
(Druce 1920!); Aqualate Mere (Druce 1925!); canal at Cheddleton
(1949!).

P. crispus L. Curled Pondweed
Canals: frequent: J, N, X, b, g, k, n–p, r–t, x–y, 15. 'Lush Pool'
(Babington 1832!).

P. x cooperi (Fryer) Fryer (*P. crispus* x *perfoliatus*)
Canal at Lichfield (Druce 1920! Edees 1948! Burges 1949!).

P. x lintonii Fryer (*P. crispus* x *friesii*)
Canal at Lichfield (Druce 1920! Foggitt 1931! Burges 1939! Edees
1948!); river Trent, Burton (Druce 1930!); marl pit near St Mary's
Church, West Bromwich (Jacobs 1948!); canal, Alrewas (Cadbury
1949!); canal, Branston (Cadbury 1949!); canal, Woodside,

Dudley (Green 1949!); canal, Tunnel Pond, Tipton (Green 1949!);
canal near Bradeshall Farm, east of Tividale, Rowley Regis
(Cadbury 1953!); canal, Wombourn (1954!).

P. pectinatus L.   Fennel Pondweed
Rivers and canals: frequent: G, W–Z, b, g, m–n, s–t, w–x, 20.
Birchills Canal near Walsall (Lowe 1834!).

ZANNICHELLIACEAE
Zannichellia palustris L.   Horned-Pondweed
Recorded for ponds, rivers and canals, mainly between Stafford,
Burton and Tamworth: rare or overlooked: (Y), (a–b), (g), (m), n,
(p), (w), 1. Yoxall, 1835 (Babington 1897). Freeford Pool near
Lichfield (1968!).

LILIACEAE
Narthecium ossifragum (L.) Huds.   Bog Asphodel
Acid moorland bogs: local: B–C, E, H, K, N, (R), W, (Z–a), (c), e,
(m), 15. Needwood Forest (Withering 1801). Plentiful on Goldsitch
Moss, the top of Gun and in the heathy field west of Swineholes
Wood: less common on Craddocks Moss and in Sherbrook Valley
and last recorded for Chartley Moss in 1886 (Masefield!).

Convallaria majalis L.   Lily-of-the-Valley
Undoubtedly native, though very local, in scree woods on the
limestone, but the early botanists also found it in other woods,
where it may sometimes have been introduced: H–J, (N), R, (Z),
(g), 4. Rough Park Wood (Gisborne 1792! and Clark 1919!);
Curborough Wood (Jackson 1837 and Moore 1889!); Manifold
Valley 'below Bincliff mines' (Garner 1844), where it is still plentiful;
Bagot's Park (Brown 1863); 'In the woods at the back of the
keeper's and near to Chartley Moss in great profusion' (Tylecote
1886); Shoul's Wood (Bagnall 1901); Hall Dale (1948); Manifold
Valley at Swainsley (1948!); Bishop's Wood, SJ/763310 (1949!).

Polygonatum odoratum (Mill.) Druce   Angular Solomon's-seal
Thickets on steep limestone hills: rare. Beeston Tor, 1879 (Ley in
BEC 1881 Rep); Dovedale (Purchas 1885); thicket below Thor's
Cave (1947!).

P. multiflorum (L.) All.   Solomon's-seal
Woods and shady banks near water: rare: C, L, (N), W, Z, (a), k,

(n), 5. The records include *P. multiflorum* x *odoratum* and most of them are probably of garden escapes. Woods at Belmont (Sneyd in Dickenson 1798). 'Near Little Haywood Bridge, Oakedge side of Trent' (Nowers 1888! Reader 1924! Edees 1957!); Oulton Coppice, Norbury (1954!); Calf Heath, SJ/925089 (1965!).

Fritillaria meleagris L.   Fritillary

Water meadows: once plentiful in a few places, but now rare. 'In a meadow near Blymhill, Staffordshire, plentifully' (Dickenson in Withering 1787). 'This very curious and rare flower adorns in great profusion some meadows about one mile from Blymhill, in the parish of Wheaton Aston' (Pitt 1794). In 1949 after a long search I found nine isolated blooms. Miss E. M. Weate tells me (in litt 1970) that the local name is 'folfalarum' and that on the first Sunday of May the children of Wheaton Aston used to go down to the 'folfer' fields to gather the flowers, often lying prone to scour the ground at grass height.

'In great abundance in a meadow on the right-hand side of the road leading from Wolseley Bridge to Stafford, not a quarter of a mile from the bridge, 7th May 1787' (Withering 1796). This was a white-flowered colony. A few plants were seen there in 1919 (Reader 1923).

'Near Tamworth opposite Mr Black's home at Mill Farm, both vars' (Power MS undated but probably about 1835). 'It has been found by me growing abundantly on an island in the Tame, near Tamworth, Staffordshire' (A.M.D. 1866). There are people still living who remember picking 'huge bunches' every year for the hospitals. There were twenty blooms in 1958. There is oral evidence that fritillaries also grew in Comberford osier-bed about seventy years ago.

'Meadows by the Dove near Uttoxeter' (Brown 1863) and 'Meadow near Doveridge Bridge' (Ridge in 1932). Other records for 'Near Abbots Bromley' (Bloxam MS); Gayton meadows (Tylecote 1885); meadow between Rugeley and Hamstall Ridware (Moore in 1903); Moddershall (1951).

Tulipa sylvestris L.   Wild Tulip

Near Statfold Hall (Garner 1844), where Brown (1863) said it was 'truly wild'. 'Near Alton, in a place once used as a marl pit' (1925). It had been known to the tenant of an adjacent cottage for fifty years. It was still there in 1955, but the cottage is in Denstone and the marl pit is now an orchard. In 1968 it was seen in a neighbouring field where houses have since been built. There is a third record for Longdon Green near Lichfield (1971). A few plants survive on a hedgebank in a field where they have been known for fifty years.

Gagea lutea (L.) Ker-Gawl. Yellow Star-of-Bethlehem
Stanton, T. S. Wilkins (1900). The plant grows with *Allium ursinum* on the south bank of Ordley Brook near Ousley Cross and has been reported for other places along the stream through Dydon Wood, but it does not flower freely and is difficult to detect. Other records for near Cheadle (1912); near Croxden (BEC 1920 Rep); near Musden Grange on the right bank of the Manifold (1950).

Ornithogalum umbellatum L. Star-of-Bethlehem
Garden escape. 'Near Bellamour House, Hill Walk, Tixall' (Clifford 1817) and again at Tixall in 1899 (1900).

Endymion non-scriptus (L.) Garcke Bluebell
Woods: A–z, 531. Needwood Forest (Gisborne 1792!).

Colchicum autumnale L. Meadow Saffron
Recorded for water meadows mainly on the eastern and southern borders of the county: (b), (d), f, (g), (r–t), (v), 1. 'Blymhill in a meadow adjacent to the Watling Street way' (Dickenson 1798). 'I have only seen it on the skirts of the eastern side of the county, but there in profusion' (Garner 1844). But Brown (1863) said it was rather scarce in the valley of the Trent. Nowers (1905) found it at Walton-on-Trent and by the Severn at Upper Arley. Fraser collected specimens from Lower Penn (1862!) and Arley (1884!). Garner (1844) quotes a record for Beaudesert but does not name his source. In 1969 G. Springthorpe sent me a clod from Beaudesert containing the plant.

Paris quadrifolia L. Herb Paris
In primrose woods and shady places about marl pits: once frequent, now rare: (F), J, (N–P), S, (U), Y–Z, (a–b), c, (d), g, (j), (n), (r–s), (v–x), 8. Near Yoxall Lodge (Gisborne 1792!) and 'Needwood Forest near Coalpit Slade' (Gisborne MS). Recent records for Chartley Moss (1950!), where it was first recorded by Tylecote (1886); Milford Hall (1953); wood at Beffcote near Gnosall (1954!); Eccleshall, pit near Highlanes Farm (1958!); pit near Standon Hall (1959!); Little Lyntus Wood near Curborough (1967); Dovedale (Smith 1871 and Shimwell 1968); Sugnall Park (1971). (Map p 238).

JUNCACEAE

Juncus squarrosus L.   Heath Rush
'It is common in mountainous, moorish and boggy places, as on the
Moorlands in Staffordshire' (Ray 1670): B–D, F–P, R–U, Y, (a),
e–f, (g), j, m, r–t, w, y, 130. (Map p 238).

J. tenuis Willd.   Slender Rush
Canal tow-paths, damp woodland rides and grassy tracks near
water: becoming frequent: C, M, Z–a, (f), n, r, t, 7. 'Sandy ground
near site of training camp, Rugeley' (Perry 1923!).

J. compressus Jacq.   Round-fruited Rush
Unconfirmed and uncertain records for Shobnall, canal bank, and
roadside between Tutbury and Burton (Brown 1863).

J. gerardii Lois.   Saltmarsh Rush
Salt marshes: very local. Kingston Pool (Garner 1844 and Reader
1923!); Branston salt marsh (Brown 1863 and Nowers 1889!);
'Site of old salt marsh near Tixall' (Reader 1923!); Shirleywich,
side of pool, SJ/985256 (1947!).

J. bufonius L.   Toad Rush
'Moist places that have been overflowed in winter' (Dickenson
1798), such as wet tracks, roadsides and pond margins: A–z, 500.

J. inflexus L.   Hard Rush
Marshy fields and damp road verges, especially in clay soils: A,
E–z, 448. Dickenson, 1798.

J. effusus L.   Soft Rush
'Marshes and miry places' (Dickenson 1798): A–z, 733. Var
compactus Hoppe is common, especially among the hills.

J. subuliflorus Drejer   Compact Rush
Common and widespread, but in smaller numbers than *J. effusus*:
A–U, W–g, j–w, y–z, 331. Dickenson, 1798.

J. subnodulosus Schrank   Blunt-flowered Rush
In marshes producing a rich flora: very local. Burton (Brown 1863);
marshy field near Yoxall Lodge (Murray 1920!); Allimore Green
Common and Dale Common (1967!).

J. acutiflorus Ehrh. ex Hoffm.   Sharp-flowered Rush
Marshy fields and wet places on heaths and moors: locally abundant: B–D, F–H, K–U, W–a, c–g, i–u, w, y, 185. Garner, 1844.

J. articulatus L.   Jointed Rush
Damp ruts and wet places by ponds and lakes: A–w, y–z, 435.
Near Yoxall Lodge (Gisborne 1791!).

J. bulbosus L.   Bulbous Rush
Acid bogs on moors and heaths: B–C, F–H, K–P, R–U, X–Z,
(a–b), e–f, m, w, y, (z), 61. Brakenhurst (Gisborne 1791!). Plants
with six stamens appear to be common but have not been separately
recorded.

Luzula pilosa (L.) Willd.   Hairy Woodrush
Woods: chiefly in the north: B–C, E, G–L, N–P, R, T–U, X, Z–a,
(b), (f), n, r, v, z, 44. 'In the wood in Callingwood Lane, April 1791'
(Gisborne MS).

L. forsteri (Sm.) DC.   Southern Woodrush
Knightley Park (Bloxam in Brown 1863). This old record has
never been confirmed and no specimen has been seen.

L. sylvatica (Huds.) Gaudin   Great Woodrush
Wooded ravines and rocky river banks: common in the Churnet
Valley and among the hills: B–D, F–J, (K), N–P, R, (S), T, Y, (a),
(f), (v), z, 45. Dingle at Cotton Hall and Burnt Wood (Dickenson
1798). (Map p 238).

L. campestris (L.) DC.   Field Woodwrush
Pastures: A–z, 605. Dickenson, 1798.

L. multiflora (Retz.) Lejeune   Heath Woodrush
Heaths, moors and damp woodland rides: common in the north
and on Cannock Chase: B–P, R–U, W, Z–a, d–f, j, m–n, s–x, z, 161.
Star Wood (Carter 1839). There are two common varieties, one
with a diffuse, the other with a compact inflorescence, but they
have not been separately recorded. (Map p 239).

AMARYLLIDACEAE

Allium vineale L.   Wild Onion
'On limestone rocks at Wetton Mill and Beeston Tor' (Garner
1844); hedgebank near Cawarden Springs Farm (Reader 1920!);
on the hillock in the car park at Wetton Mill (1946); West
Bromwich, railway embankment (1949); abundant on a roadside
bank between Kinver Bridge and Dunsley Dene (1960!); Hanford,
Stoke, in a meadow near the confluence of the Lyme Brook and the
Trent (1960!). Both var *vineale* and var *compactum* (Thuill.) Boreau
occur.

A. oleraceum L.   Field Garlic

'On a rock in Wetton Valley; in St Chad's churchyard, Lichfield' (Garner 1844); Dovedale, in a rock fissure at about 1,100ft (Purchas 1885); near Wetton Mill (Nowers 1892!) and Sugarloaf (1947!).

A. ursinum L.   Ramsons

By streams in woods, especially on limestone and marl: locally abundant: B–C, E, (G), H–L, N–U, W–X, Z–b, d, f–g, j–k, n, r, t–w, y–z, 129. 'Meadows at Horsebrook; near Marchington Woodlands; Ilam' (Pitt 1794). (Map p 239).

Galanthus nivalis L.   Snowdrop

Naturalised by streams in woods, often near houses, but sometimes outlasting the houses: incompletely recorded: (F), J, (K), (M–N), P, (U), Y, (b), (g), (j), (n), 3. 'In orchards, a beautiful early flower of welcome appearance' (Pitt 1794). Manifold Valley near Castern, several patches in damp places in the wood, looking quite wild (1951!). Garner recorded them for Castern in 1844.

Narcissus pseudonarcissus L.   Wild Daffodil

Local in meadows and woods, especially by streams, and in church-yards: C, G, K–P, R, T–U, W–Z, (a), f, (g), (k), n, r, (x), 25. Pitt, 1794. 'Orchards and hedges near houses, rare; covers more than an acre in a field near Willowbridge Lodge; and grows very profusely in an adjacent wood in the parish of Muccleston' (Dickenson 1798). 'Abundant in rich fields' (Garner 1844). Perhaps introduced in some places, but often evidently native. Where it appears quite native in churchyards, as at Fradswell, Gratwich and Kingston, it can be found sparingly in neighbouring fields. But the plough is rapidly destroying it in acres unprotected by the churchyard wall. It survives in woods near Mucklestone and is plentiful at Longdon, though ravaged annually by unscrupulous pickers. In the north most of the localities are fields adjoining farm houses, but there is no proof of introduction. The usual story is that the daffodils have been growing in the field for as long as any one can remember and that there were more of them in former days than there are now.

IRIDACEAE

Iris pseudacorus L.   Yellow Iris

Marshy places in woods and by the sides of rivers, lakes and canals: C, E–U, W–y, 202. Pitt, 1794. (Map p 239).

Crocus nudiflorus Sm.   Autumn Crocus
Recorded for orchards, churchyards and fields near houses:
frequent in the north between Eccleshall and Leek: (F), G, L–M,
(N), S–T, (j), 10. Wolstanton (Butt in Clifford 1817). This species
has a long history at Wolstanton. According to Garner (1844) it
grew 'abundantly in a field near Wolstanton and in two other
fields near'. In 1876 Elizabeth Edwards (Jour Bot) said that it had
existed at Wolstanton for a century in plenty, 'in a hilly pasture-
field just below Wolstanton church'. But by 1926 it was thought to
be extinct (Ridge in Flora). It was rediscovered in 1952 and is now
again thought to be extinct.

C. purpureus Weston   Spring Crocus
'Occasionally in the Trent meadows near Burton' (Brown 1863);
Tutbury Road, near Horninglow (Burton Flora 1901).

DIOSCOREACEAE

Tamus communis L.   Black Bryony
'Hedges and thickets' (Pitt 1794): H–w, y–z, 315. (Map p 239).

ORCHIDACEAE

Epipactis palustris (L.) Crantz   Marsh Helleborine
Rich marshy fields: rare. 'Moors near Moreton' (Dickenson
1798); 'Meadow at foot of Barr Beacon' (Ick in Garner 1844);
Fairoak (Garner 1844); Stanton, SK/1247 (1938!); Allimore
Green Common (1967).

E. helleborine (L.) Crantz   Broad-leaved Helleborine
Woodland rides, wood margins and shady roadside banks: frequent,
but usually only a few plants grow together: B, F–H, K–L, N–P,
R, (T), X, (Y), a, (d), f, j–k, (m), r, w, y, 32. 'In a lane from Oulton
Heath towards Barlaston' (Forster 1796).

Spiranthes spiralis (L.) Chevall.   Autumn Lady's-tresses
Unconfirmed records for meadows at Kingswinford (Bree in
Purton 1821) and Dovedale (Smith 1871).

Listera ovata (L.) R.Br.   Common Twayblade
Common on the Carboniferous limestone in old quarries, shallow
roadside ditches and at the foot of grassy slopes: less common
elsewhere in damp woods, especially on the Keuper marl: B–C,
G–L, N–P, R, W–X, (Y–Z), d, f–g, i–j, (r–t), v–w, 43. 'Bushy pit at
Buttermilk Hill, 1792' (Gisborne MS).

Neottia nidus-avis (L.) Rich. Birdsnest Orchid
Unconfirmed records for Henhurst Wood (Brown 1863); Dovedale (Smith 1871); Cotton Valley (Yates in 1876); near Penn (Bagnall 1888!); Manifold Valley (Bagnall 1901).

Hammarbya paludosa (L.) Kuntze Bog Orchid
'Norton Bog, Cannock Wood' (Bagot in Withering 1801).

Coeloglossum viride (L.) Hartm. Frog Orchid
'Turfy hills' (Smith 1871), mainly on the limestone: local: (B–C), (G), H–J, (M), N–P, (R–U), (b), (d), (w), 7. 'Blymhill, in the moor, at the Restlars' (Dickenson 1798). Often plentiful in Caldonlow quarries.

Gymnadenia conopsea (L.) R.Br. Fragrant Orchid
Limestone hillsides and wet pastures: local: (G), H, (J), (N), P, d, (m), (t), 3. Near Blymhill and near Cotton (Dickenson 1798). Recent records for Stanton (1933!); Manifold Valley, hill north of Ecton and among the rocks by the roadside (1944!); Allimore Green Common (1967). Subsp *densiflora* (Wahlenb.) G. Camus, Bergon and A. Camus: Stanton in 1968 (Miss Hollick in litt).

Pseudorchis albida (L.) A. and D. Löve Small-white Orchid
One unconfirmed and doubtful record for a field near Hollington in 1880 (Edwardes in Goodall 1882).

Platanthera chlorantha (Custer) Reichb. Greater Butterfly-orchid
Limestone hillsides, grassy fields, railway embankments, heathy pastures and woods: local in the north: C, (F), G–H, (J), K, N–P, (a–b), 15. 'Callingwood Lane by the wood, 1791' (Gisborne MS), probably this species. In several places near Stanton and on Ecton Hill.

P. bifolia (L.) Rich. Lesser Butterfly-orchid
Alstonfield, 'in some plenty on one part of the half-reclaimed Staffordshire moorland in this parish' (Purchas 1885); Highfield near Alton (Berrisford 1904!); Black Heath (1938!); Stanton and Reaps Moor in 1947.

Ophrys apifera Huds. Bee Orchid
Yoxall Lodge, very rare (Churchill Babington in Watson 1837); Wren's Nest (Garner 1844); near Alton (1899); wood at King's Bromley (Reader 1922!); Mons Hill, one plant (1950); Daw End near Walsall, two plants (1950).

O. insectifera L.   Fly Orchid
Dovedale, Staffordshire (Smith 1871); Manifold Valley (1934!),
one or two plants are seen from time to time by the side of the old
railway track near Beeston Tor.

Orchis ustulata L.   Burnt Orchid
'Turfy open hills' (Smith 1871). Kingswinford (Bree in Purton
1821); Weaver Hills (Berrisford 1909!).

O. morio L.   Green-winged Orchid
Rich moist meadows: once frequent, now very rare: R, (W–X),
(a–b), (f–g), (m), (t), (v–w), 1. 'Bushy pit near Buttermilk Hill,
1792' (Gisborne MS). 'Betwixt Lynn and Chesterfield near Lichfield'
(Power MS undated). Dickenson (1798) said it was common and so
did Garner (1844), though less common than *O. mascula*. But there
are so far only three records for the twentieth century: 'Meadow
between North Longdon and Beaudesert' (Clark 1920!); Trent
meadows near Yoxall (1924); Eccleshall, rough pasture at
Greatwood, SJ/785305, one plant (1960).

O. mascula (L.) L.   Early-purple Orchid
Grassy limestone hillsides, woods on the Keuper marl and railway
embankments: common in the Manifold Valley and Dovedale,
local or infrequent elsewhere: (F), H–K, N–P, R, W–X, (Y–Z), a,
d–e, g, (j), n, (s–t), 28. 'Wood in Callingwood Lane, most abundantly, April 1791' (Gisborne MS).

Dactylorhiza fuchsii (Druce) Soó   Common Spotted-orchid
Hillsides and old abandoned quarries on the limestone, marshy
fields and railway embankments, our commonest orchid: A–C,
G–P, R–U, W–a, c–g, j–n, z, 109. 'The *Orchis maculata*, or spotted
orchis, grows so profusely in many parts of this country, that the
walls of cottages are frequently decorated with garlands composed
of its purple spikes' (Gisborne 1797). Plants with pure white
flowers are found occasionally. (Map p 240).

D. maculata (L.) Soó subsp *ericetorum* (E.F.Linton) Hunt and
   Summerh.   Heath Spotted-orchid
Heaths and moors in damp acid soil: probably frequent in the
north: C, H, K, P, R, X, d, 14. Dovedale (Thornton 1924!).

D. incarnata (L.) Soó   Early Marsh-orchid
Sparingly on Loynton Moss (1954), but perhaps doubtful: no
specimen was kept and it has not been seen again.

D. praetermissa (Druce) Soó　　　　　　Southern Marsh-orchid
Wet rushy meadows, peaty swamps and reed beds by canals and
meres: frequent in the south, but seldom in quantity: R–T, W, Y,
a, c–d, f, 16. 'Marshy meadows, common' (Dickenson 1798), as
Orchis latifolia but probably this species. Yoxall, marshy ground by
Lin Brook along Foxholes Lane (1956!), where Gisborne found
O. latifolia (MS undated). Near Patshull (Lady Joan Legge in
BEC 1917 Rep), det Druce. Perhaps most plentiful today in
marshy meadows about Aqualate Mere (1953!), but it can also be
seen on Allimore Green Common and in the fritillary field at
Wheaton Aston.

D. purpurella (T. and T. A. Stephenson) Soó
　　　　　　　　　　　　　　　　Northern Marsh-orchid
Very local in the north in wet peat and sand. Moss Carr, R. H.
Hall (1946!); Trentham gravel pit (1954!); Grindon Moor in 1957,
half a dozen plants in a roadside ditch. The orchids survive at
Trentham.

Anacamptis pyramidalis (L.) Rich.　　　　　　Pyramidal Orchid
Plantation near Uttoxeter (Hewgill in Garner 1844 and Bagnall
1878!); Catholme Lane, Barton (Hewgill in Garner 1844); Manifold
Valley (Fraser 1864); Wetton, rare (Smith 1871). Miss C. Roscoe,
who lived at Back of Ecton for many years, told me in 1948 that she
used to see it by the roadside at Apes Tor.

ARACEAE

Acorus calamus L.　　　　　　　　　　　　　Sweet-flag
Established in a few places at the edges of rivers, lakes and canals,
but originally introduced: (K), L, (M), (R–S), Z–b, e, (f), (h), k, n, 9.
'River at Tamworth at the bottom of Mr Oldershaw's garden'
(Withering 1787).

Arum maculatum L.　　　　　　　　　　　Lords-and-ladies
Woods and shady hedgebanks, especially on limestone and marl:
E–F, H–L, N–z, 378. Pitt, 1794. (Map p 240).

LEMNACEAE

Lemna polyrhiza L.　　　　　　　　　　Greater Duckweed
Ponds and canals: frequent in the centre of the county: (K–L), S,
(T), W–Y, f, (j), (t), 6. Pit by the roadside between Kiddemore
Green and Bishop's Wood (Dickenson 1798).

L. trisulca L. Ivy-leaved Duckweed
Ponds and canals: widespread and locally plentiful: B, G, K, M–N,
S–U, W–Z, (b), d–r, t, x–y, 72. Blymhill (Dickenson 1798).

L. minor L. Common Duckweed
Still waters: B, E–u, w–y, 438. Dickenson, 1798.

L. gibba L. Fat Duckweed
Ponds and canals: frequent in the south: W–Y, (d–e), f–h, k–m,
(n), p, s–t, x, 18. 'Blymhill Lawn, in the pit nearest the High Hall'
(Dickenson 1798).

SPARGANIACEAE

Sparganium erectum L. Branched Bur-reed
Margins of canals, ponds and lakes: B, E–P, R–u, w–y, 275. Pitt,
1794. The subspecies have not been studied.

S. emersum Rehm. Unbranched Bur-reed
Frequent in canals and also recorded for pools and marshes:
(F), G, K, (L), (N), P, (T), (W), (Z), (b–c), (f–g), n, (r), s, (w), 9.
'Sides of pools . . . in shallow water in Pitmoor Pool, Weston-under-
Lizard' (Dickenson 1798).

S. minimum Wallr. Least Bur-reed
Ditches in Bagot's Park (Brown 1863); Enville Common (Fraser
1865!); Codsall (Fraser 1874! in hb Bagnall); Callingwood (Nowers
1896!).

TYPHACEAE

Typha latifolia L. Bulrush
Reed swamps round lakes and ponds: B, E–H, K–N, Q–u, w–x,
221. Pitt, 1794.

T. angustifolia L. Lesser Bulrush
Ponds and lakes: uncommon: K, L, (W), X–Y, (Z–b), h, i, (m),
(r), t, 8. 'Pool near Chartley House' (Bagot in Withering 1801).
Brown (1863) said it was 'frequent in ditches and railway cuttings
round the borders of Needwood Forest'. Is this true today? It grew
in one of the lily pits at Branston (1956!).

CYPERACEAE

Eriophorum angustifolium Honck. Common Cottongrass
Acid peat bogs: less common in the south than formerly, but

abundant in wet places on the northern moors: B–C, E–H, K–P, R–T, W, Z–a, d–e, (f), m–n, (r), t, 52. 'Aqualate Mere . . . covering several acres' (Withering 1787). (Map p 240).

E. latifolium Hoppe   Broad-leaved Cottongrass
'Willowbridge and Chartley Moss' (Garner 1844), uncertain, but a record for Penn Common (Fraser 1883! and 1884!) is correct.

E. vaginatum L.   Harestail Cottongrass
Wet peaty places: common on the moors: B–C, E–H, K,M–N, R–S, (W), Z, (c), e–f, (g), (m), t, (z), 35. 'Generally grows in wet places among stones' (Pitt 1796, p 200).

Scirpus cespitosus L.   Deergrass
Damp acid moors: local in the north: B–C, (F), H–K, M–N, (R), (Y), (d), (f), (m), 10. 'Weston-under-Lizard, in heathy land' (Dickenson 1798). Carter (1839) said it was 'extremely common and abundant'. Maer Heath (Daltry 1840!). Often recorded for Norton Bog and said to be abundant there before the development of coal mining (Bagnall 1887!). Recent records for Archford Moor, Black Heath, Cloud Side, Craddocks Moss, Gun, Swallow Moss and Wetley Moor.

S. maritimus L.   Sea Clubrush
Salt marshes: rare. Ditch or stream by Kingston Pool (Garner 1844, Fraser 1865! Reader 1923!); Branston meadows (Brown 1863, Nowers 1889!) and Branston gravel pits, SK/222206, G. A. Arnold (1959!); Great Haywood, north side of canal near Swivel Bridge west of Haywood Mill (Reader 1923! Edees 1947!).

S. sylvaticus L.   Wood Clubrush
Shady stream sides and marshy places in woods: frequent: (B), C, G–H, K–L, N–T, W, (Y), Z, (b), c–d, f, j, m–n, r, t, v, y, 40. 'Weston-under-Lizard, in bogs near the mill' (Dickenson 1798).

S. lacustris L.   Common Clubrush
'Rivers and lakes' (Dickenson 1798), making tall stands in the water: frequent in the south: R, W–Y, b, d–u, w, 40.

S. tabernaemontani C.C.Gmel.   Grey Clubrush
The salt marsh records are the most reliable. Shirleywich (Garner 1844); Branston salt marsh (Nowers 1889! and Edees 1956!); in a marshy hollow at Shirleywich (Reader 1924).

S. setaceus L. Bristle Clubrush
Marshy meadows and damp sandy places: frequent: B, (F), G, (K), L–P, R, (S), T–U, (V), W–X, (Y), a–g, (m), r–t, y–z, 35. 'Blymhill in wet sandy land' (Dickenson 1798).

S. fluitans L. Floating Clubrush
Recorded for bogs and shallow waters in past days, but not seen recently. Whitemere Bog, 1792 (Gisborne MS); bog near White Sitch Pool (Dickenson 1798); Craddocks Moss (Garner 1844); Cannock (Fraser 1864!); Sherbrook Valley (Bagnall 1901); Norton Bog (Bagnall 1901 and Reader 1923!).

Eleocharis acicularis (L.) Roem. and Schult. Needle Spikerush
One unconfirmed record for Sherbrook Valley (Bagnall 1901).

E. quinqueflora (F. X. Hartmann) Schwarz
Few-flowered Spikerush
Chartley Moss (Brown 1863); Brocton, bog in Oldacre Valley (1965!).

E. multicaulis (Sm.) Sm. Many-stalked Spikerush
'Marsh betwixt Shugborough and Brocton' (Power MS undated). There is a correctly named specimen (also undated) in Power's herbarium with these words against it, but, as a Warwickshire locality appears in faded ink at the foot of the page, we cannot be sure that the plant was gathered in Staffordshire.

E. palustris (L.) Roem. and Schult. Common Spikerush
Marshy fields and at the edges of ponds and lakes: B, F–G, J–P, R–U, W–t, w, y, 157. Near Stafford (Stokes in Withering 1787).

Blysmus compressus (L.) Panz. ex Link Flatsedge
Field opposite Yoxall Lodge (Babington in Watson 1837); Warwickshire Moor near Tamworth, G. A. Arnold (1962!).

Schoenus nigricans L. Black Bogrush
Unconfirmed records for Moreton Moors (Dickenson 1798) and Norton Bog (Fraser 1864).

Rhynchospora alba (L.) Vahl White Beaksedge
Unconfirmed records for Chartley Moss (Bagot in Clifford 1817) and Craddocks Moss (Garner 1844 and Fraser 1864!).

Cladium mariscus (L.) Pohl Great Fensedge
Fen carr: very rare. Moors near Moreton (Dickenson 1798);

Chartley Moss (Howitt in Watson 1837, Reader 1924! and R. A. Boniface in 1951).

Carex laevigata Sm.   Smooth-stalked Sedge
Marshy woods and by moorland streams in the north: local: C, F–H, K, n, 11. Madeley, H. W. Daltry (Ridge in Flora). Canal side between Consall Forge and Cheddleton (1937!) and several other places near Cheddleton; between Swainsmoor and the Traveller's Rest, several records; Walton's Wood (1942!); wood on east side of Knypersley Pool (1955); Hovel Covert near Weeford (1957).

C. distans L.   Distant Sedge
Salt marsh near Weston-upon-Trent (Thornton 1923!).

C. hostiana DC.   Tawny Sedge
Heaths and commons in wet peaty soil, in two places over limestone and always associated with a rich flora: local. Penn Common (Fraser 1884! and again in 1963); Stanton, SK/1247 (1933!); roadside between Calton Moor and Blore (1947!); bog near Brocton (1948!); Barlaston Common (1955!); Whittington Heath (1962); Allimore Green Common (1967). *C. hostiana* x *demissa:* Stanton, E. Nelmes (1947!).

C. binervis Sm.   Green-ribbed Sedge
Frequent in damp and also in apparently dry places on acid heaths and moors: (A), B–C, G–H, L, (M), N–P, R, T–U, Z, (a), (e), (m), (r), t, 23. Cheadle (Carter 1839).

C. demissa Hornem.   Common Yellow-sedge
Marshy fields in the lowlands and by streams on the moors: frequent: C, F–H, K–L, N–P, R–T, W–X, Z–a, (b), c, e–f, (m), n, (s), z, 46. Near Yoxall Lodge (Gisborne 1791!).

C. sylvatica Huds.   Wood-sedge
Woods, especially on the Keuper marl: G–L, (M), N–S, (V), W–X, Z–b, d, f, r, (t), v–w, z, 50. 'Blymhill in a hag or grove at the Restlars' (Dickenson 1798).

C. pseudocyperus L.   Cyperus Sedge
Marl pits and at the edges of drains and stagnant pools in woods: frequent in the south: E, K–L, R–S, W–a, (b), c–f, k–p, (r), w, y, 32. Whitmore and Ashley (Garner 1844).

C. rostrata Stokes   Bottle Sedge
Peat swamps and at its best when growing in shallow water at the edge of a lake: local: B–C, (F), G–H, K–M, (N), R–U, W, Y–Z (b), d–f, j, m, r, t, w, 42. Belmont and Stoke meadows (Garner 1844). (Map p 240).

C. vesicaria L.   Bladder-sedge
Marshy borders of lakes and ponds: local: (K), R–S, W, Y–Z, (f), g, r, w, 11. Near Pool Hall, Lower Penn (Fraser 1863! and again in 1961!); Black Lake, Trentham Park (1937!); Foucher's Pool (1954!); Dunstal Pool, Bagot's Park (1956!); Little Gorse near Creswell (1959); Loynton Moss (1966!) and several places between there and Eccleshall. Dickenson (1798) records this species for Moreton Moors, Pitmoor Pool and ditches near Aqualate Mere, but he may have intended *C. rostrata*.

C. riparia Curt.   Greater Pond-sedge
Margins of lakes, rivers and canals, particularly between Burton and Tamworth: G, K, R–S, W–Z, b, d–e, (f), g–j, (m), n–p, (r), t–u, y, 37. Dickenson, 1798.

C. acutiformis Ehrh.   Lesser Pond-sedge
Marshy fields, ditches, ponds and by the sides of rivers and canals: common: B–C, G–H, K–N, Q–U, W–r, t–u, 152. Dickenson, 1798.

C. pendula Huds.   Pendulous Sedge
Sometimes planted, but native in wet woods in stiff soils: locally plentiful, especially in the rich woods near Arley: B, K–L, N, (T), Z–a, (b), (f), j, n, t, v–w, z, 19. In 1961 it was found to be abundant in the wood where it was first recorded for the county: 'This most elegant and rare plant adorns, in the greatest profusion, the edges of a rivulet in the Big-Hide-Rough wood, near Brewood' (Dickenson 1798).

C. strigosa Huds.   Thin-spiked Wood-sedge
By streams in woods: rare. Anslow and Tatenhill (Brown 1863); Arley Dingle (Fraser 1883! and again in 1954!); Bagot's Wood (Reader 1918!); Springpool Wood and Whitmore Wood (1952!); Forest Banks and Knightley Park (1956!); Eccleshall, with *C. sylvatica* between the pools in Sugnall Park, SJ/801301 (1961!).

C. pallescens L.   Pale Sedge
Woodland rides, damp hayfields bordering woods, grassy lake

sides: frequent, but usually only a few tufts are seen in any one place: C, G, J–K, (M), N–P, R–T, W–X, Z–a, d, (f), k–m, (n), z, 24. 'In the wood near Rough Park' (Gisborne 1791!).

C. panicea L.   Carnation Sedge
Wet meadows and moorland bogs: common: B–C, E–P, R–U, W–a, c–g, j–n, r–t, y–z, 168. Blymhill (Dickenson 1798).

C. limosa L.   Bog-sedge
'Moreton Moors: bogs at Pitmoor Pool, Weston-under-Lizard' (Dickenson 1798). Bagnall (1901) thinks that a variety of *C. flacca* was mistaken for this species, but Dickenson was a careful botanist.

C. flacca Schreb.   Glaucous Sedge
Water meadows on basic soils, canal tow-paths, lake-sides, woods, road verges, railway embankments and limestone grassland: common: B–C, E–P, R–U, W–a, d–g, i–n, r–w, y–z, 178. 'In Thatchmoor meadows' (Gisborne 1791!).

C. hirta L.   Hairy Sedge
Common in damp grassy places: E–w, y, 257. Blymhill (Dickenson 1798).

C. pilulifera L.   Pill Sedge
Dry heaths, woodland rides, sandy roadsides and the drier parts of peat mosses: frequent, especially on Cannock Chase and in the Churnet Valley: (B), C, F–K, M–P, R–S, U, Z, (d), e–f, (m), v–w, z, 32. 'Blymhill Far Restlar's Meadow' (Dickenson 1798).

C. caryophyllea Latourr.   Spring-sedge
Dry well-drained grassy slopes, especially in calcareous soils, and more rarely in marshy fields: common on the Carboniferous limestone and frequent between Madeley and Forton in the west: C, (F–G), H–L, N–P, R–S, W–X, a, (f), (m), (s), v, z, 34. Dickenson, 1798. (Map p 241).

C. elata All.   Tufted-sedge
Ditches and ponds, growing in the water: local on the western side of the county. 'In Pitmoor Pool, by the side of the dam, and in various parts of the pool in large tufts, Weston-under-Lizard' (Dickenson 1798); Shelmore Wood (Bagnall 1896! and C. E. Andrews in 1953); Chapman's Wood south of Moreton (1954!); ditch on west side of Aqualate Mere (1954!); Windsend near Eccleshall (1959!); Balterley Heath (1961); Loynton Moss (1966!).

C. acuta L.   Slender Tufted-sedge

Lakesides and river banks: local: (K), S, (b), (f), g–h, (m), p, (t), (z), 4. Dickenson, 1798. Several places by the Trent between Burton and Tamworth; Walton's Wood (Daltry 1917!); Mavesyn Ridware (Reader 1922!); Trentham (1934!).

C. nigra (L.) Reichard   Common Sedge

Common in marshy fields and locally abundant in wet places on the moors where streams percolate through the peat: B–C, E–P, R–U, W–a, (b), c–g, j–p, s–t, w, y–z, 230. Dickenson, 1798.

C. paniculata L.   Greater Tussock-sedge

Canals, alder woods and peat swamps on the margins of lakes: common: B, E–H, K–N, Q–T, W–Z, (a), b–g, i–r, t–u, w, y, 105. 'In Thatchmoor meadows' (Gisborne 1791!).

C. diandra Schrank   Lesser Tussock-sedge

Marsh near Oulton (Bagnall 1901); south side of Podmore Pool (1959!).

C. otrubae Podp.   False Fox-sedge

Ditches and canal sides: common in the south: L, (N), Q, S–u, w–x, 148. Dickenson, 1798. (Map p 241).

C. disticha Huds.   Brown Sedge

Marshy fields: local: K, S–T, W, Y–Z, (b), c–e, (f), g, j–p, t–u, 45. 'Blymhill, in peaty meadows, frequent' (Dickenson 1798). (Map p 241).

C. divulsa Stokes   Grey Sedge

Shady hedgebanks on the Carboniferous limestone and Keuper marl: rare. Anslow (1945!) and other places in Needwood Forest; Waterhouses (1954). Earlier records uncertain.

C. spicata Huds.   Spiked Sedge

Frequent on the Keuper marl, where it is nearly always found in small quantity in marshy fields, but it has also been recorded for the Carboniferous limestone and grows occasionally on hedge-banks and by roadsides: J, L, N–P, S–T, V–b, d, y, 29. The early botanists did not distinguish this species from *C. muricata*, but the first two words of the earliest record, 'Moist places and also dry ditch banks' (Dickenson 1798) point to *C. spicata*. All the twenty-seven local specimens in my herbarium showed plainly, when freshly gathered, the red colouring of the scales and bracts, which Nelmes used to say was a good distinguishing character.

C. muricata L. Prickly Sedge
Frequent throughout most of lowland Staffordshire on dry roadside banks: G, M, P–S, U, W–r, t, w, z, 83. Hawkesyard Park (Reader 1917!).

C. elongata L. Elongated Sedge
Very local in fen woods. Balterley Heath (1961!); Loynton Moss (1966!).

C. echinata Murr. Star Sedge
Wet peaty places, especially among the hills and on Cannock Chase: C, (F), G–H, K, M, (N), P, R–T, (U), Y–Z, d–f, j, m–n, (r), s, (t), (w), x, 55. Near Yoxall Lodge (Gisborne 1791!). (Map p 241).

C. remota L. Remote Sedge
Shady roadside ditches and damp muddy places in woods: common in the Churnet Valley, Needwood Forest and the west of the county: B, E, (F), G–H, K–L, N–U, W–a, (b), c–n, r, (t), v–w, y–z, 112. Dickenson, 1798. (Map p 242).

C. curta Gooden. White Sedge
Local in acid marshes bordering lakes and in the wettest parts of peat bogs, often in drainage ditches and rills of running water: C, F–H, K, (L–N), R–S, (W), (Z), (c), (e), (j), (m), (w), 19. 'Boggy ground near Aqualate Mere; by the side of Pitmoor Pool; Weston-under-Lizard; in the Big-Hide-Rough near Brewood' (Dickenson 1798). Half the recent records are for the moorlands between Leek and Buxton.

C. ovalis Gooden. Oval Sedge
Water-logged pastures and wet places on heaths and moors: A–U, W–g, j–p, s–v, x–z, 266. Dickenson, 1798.

C. pulicaris L. Flea Sedge
Frequent in wet places over limestone, less common in boggy places on heaths and by moorland streams: C, H, (J), N–P, R–S, W, (Z–a), d, e, m, (z), 14. 'Scotch Plain and the Hollyfalls Bog, 1791' (Gisborne MS).

C. dioica L. Dioecious Sedge
'Weston-under-Lizard, in boggy ground near the mill' (Dickenson

1798); Chartley (Brown 1863); Sherbrook Valley (Bagnall 1901); Brocton, in a small bog in Oldacre Valley (1965!).

GRAMINEAE

Phragmites australis (Cav.) Trin. ex Steud.   Common Reed
'Rivers, lakes and ditch-banks, in wet, marshy land' (Dickenson 1798): G, J–L, Q–S, W–d, f–s, u, 53. Pitt, 1794. (Map p 242).

Molinia caerulea (L.) Moench   Purple Moor-grass
Damp or wet places on heaths and moors: locally dominant: B–C, E–P, R–U, W–a, d–f, j–n, r–t, w, y–z, 115. 'Brakenhurst little bog and about it, July 1972' (Gisborne MS). (Map p 242).

Sieglingia decumbens (L.) Bernh.   Heath-grass
Rough heathy pastures: B–D, F–P, R–U, W–a, (b), d–g, k–t, w, z, 121. Dickenson, 1798.

Glyceria fluitans (L.) R.Br.   Floating Sweet-grass
Wet meadows: B–u, w–y. 363. 'Generally growing in water, it is a sweet and good herbage and very productive. Many a poor old horse has been bogged in searching for this grass, of which they are remarkably fond' (Pitt 1794). *G. fluitans* x *plicata* occurs frequently.

G. plicata Fr.   Plicate Sweet-grass
Wet fields and by the sides of ponds and ditches, often in shallow slowly moving water: A–C, E–w, y, 272. Brown, 1863.

G. declinata Bréb.   Small Sweet-grass
In wet mud at the edges of ponds: B–U, W–t, w, y, 216. Purchas (1885) was probably the first to recognise this species in the county: 'There are two forms of *G. plicata*, one . . . is glaucous, with blunt broad leaves.' *G. declinata* x *fluitans:* pond near Thorswood, Stanton, H. K. Airy Shaw, 1947, det. C. E. Hubbard (1948).

G. maxima (Hartm.) Holmberg   Reed Sweet-grass
Ponds and riversides, but particularly common along the canals: F–G, L–M, (N), Q, S–b, d–u, w–y, 176. 'Very common by the sides of streams, in hedges, and sometimes in pretty broad patches on water meadows' (Pitt 1794) and again: 'Sides of rivers, six feet high, it frequently occurs on the banks of the canals near Newcastle' (Pitt 1817). (Map p 242).

Festuca pratensis Huds.   Meadow Fescue
Damp grassy places: B–P, R–z, 374. Pitt, 1794.

F. arundinacea Schreb. Tall Fescue
Wet meadows, river banks and canal sides: frequent: G–J, Q–S, W–X, b, d–e, (f), h–r, t, w–x, 40. Hawkesyard Park (Reader 1917!). Earlier records uncertain.

F. gigantea (L.) Vill. Giant Fescue
Damp woods and shady hedgebanks, especially on the limestone and Keuper marl: A–D, F–k, n–y, 268. Dickenson, 1798. (Map p 243).

F. altissima All. Wood Fescue
Dovedale Wood (Shimwell 1968).

F. rubra L. Red Fescue
Pastures, hayfields, roadsides: B–C, E–T, W–u, w–y, 233. Walls of Dudley Castle (Withering 1796).

F. ovina L. Sheep's-fescue
Dry grassy places: B–U, W–z, 394. Dickenson, 1798. A glaucous form is common on the limestone.

F. tenuifolia Sibth. Fine-leaved Sheep's-fescue
This may prove to be common on the Bunter sandstone, but at present there are only seven records: Cannock Chase (Fraser 1879!); Norton Bog (Bagnall 1897!); Barr Beacon (Bagnall 1901); 'Moor in Staffordshire' (BEC 1920 Rep); Lichfield (1945!); Chorlton Moss, field between wood and railway (1949!); golf course, Harborne (1959!).

Lolium perenne L. Perennial Rye-grass
Common everywhere as an introduced or native plant: A–z, 795. Pitt, 1794. 'Ray-grass, called universally, though corruptly, rye-grass in Staffordshire. Cultivated and sown, mixed with clover, to great advantage in stiff land' (Dickenson 1798). *Festuca pratensis* x *L. perenne* occurs frequently.

L. multiflorum Lam. Italian Rye-grass
Naturalised in many places: A–C, E–w, y, 377. Brown, 1863.

L. temulentum L. Darnel
Formerly in cornfields: today a rare casual. Pitt, 1794. 'Darnel . . . a very pernicious weed found only amongst corn' (Dickenson 1798); Burton (Nowers 1904! and Druce in BEC 1929 Rep); Oakamoor (Berrisford 1906!); Newcastle, bird-seed alien (1970).

Vulpia bromoides (L.) Gray   Squirrel-tail Fescue
Dry sandy places, gravel pits, walls, cinder tracks and railway lines:
frequent, especially in the west: E, J–L, Q–S, W–Z, c, e–f, n, r, t, w,
y–z, 50. Dickenson, 1789. (Map p 243).

V. myuros (L.) C.C.Gmel.   Ratstail Fescue
Dry gravelly places: rare. 'On the road leading from Blymhill
to Shrewsbury' (Dickenson in Withering 1796); Kinver, 1862
(Mathews 1884); Burton (Curtis 1930); Brewood (1962); Uttoxeter,
between railway lines in station goods yard (1965).

Puccinellia maritima (Huds.) Parl.   Common Saltmarsh-grass
Remains of old salt marsh near Tixall (Reader 1923!); Pasturefields, SJ/992248 (1948!).

P. distans (L.) Parl.   Reflexed Saltmarsh-grass
Shirleywich (Reader 1923!); Stafford, SJ/923259 (1961!).

P. rupestris (With.) Fernald and Weatherby   Stiff Saltmarsh-grass
All we know about this species in Staffordshire is contained in the
following three statements: 'Muddy salt marsh (near brine springs)
between Stafford and Baswich' (Reader 1923!); 'In profusion in
two salt marshes near Stafford and Baswich' (Reader 1924); 'By
the river about two miles from Stafford (brine baths are near)'
(Talbot in BEC 1929 Rep).

Catapodium rigidum (L.) C.E.Hubbard   Fern-grass
Walls and limestone rocks: apparently rare, but perhaps overlooked. Tutbury Castle (Dickenson 1798), seen there in 1956;
Oaks Plantation, Shobnall (Brown 1863); Dovedale (Fraser
1879!); Mavesyn Ridware (Reader in Bagnall 1901); Bunster Hill
(1946!); Hall Dale (1947!).

Poa annua L.   Annual Meadow-grass
Waysides and waste places: A–z, 784. 'Needwood Forest, chiefly
near roadsides' (Pitt 1794). 'Footpaths and gravel walks, from
which it is very difficult to extirpate it' (Dickenson 1798).

P. nemoralis L.   Wood Meadow-grass
Woods and steep shady roadside banks: frequent: B, G, K–L,
(M–P), R–U, W–Y, a–c, e, (f), g, j, q, (r), s, (t), v–w, y–z, 41.
Big Hyde Rough near Brewood and Belmont (Dickenson 1798).

P. compressa L.   Flattened Meadow-grass
Walls and disused railway tracks: rare. Abbey walls, Burton

(Brown 1863); wall top, Hill Ridware (Reader 1920!); bridge over canal, Forton (1944!); old railway track, Waterhouses (1955!); Willingsworth furnaces near Wednesbury (1957); Croxden Abbey walls (1958); Rushton railway station (1965!).

P. pratensis L.   Smooth Meadow-grass
In many habitats: B–z, 567. Pitt, 1794.

P. trivialis L.   Rough Meadow-grass
Meadows, cultivated fields, woods and waste ground, in damp or wet places: A–z, 670. Pitt, 1794.

P. chaixii Vill.   Broad-leaved Meadow-grass
Patshull (Druce in BEC 1917 Rep).

Catabrosa aquatica (L.) Beauv.   Whorl-grass
Wet places by streams, often in shallow slowly flowing water: frequent: K, M, P–R, (X), Y–Z, (b), c–g, m–n, (r), t, 30. 'In shallow waters, Blymhill' (Dickenson 1798).

Dactylis glomerata L.   Cocksfoot
Fields, roadsides and waste places: A–z, 796. Pitt, 1794.

Cynosurus cristatus L.   Crested Dogstail
Meadows and pastures: A–z, 732. Pitt, 1794.

C. echinatus L.   Rough Dogstail
A rare casual in gardens and on waste ground. Oakamoor (Berrisford 1902!); weed in garden, Hawkesyard (Reader 1922!); Tame Valley Bridge (BEC 1926 Rep); Burton (BEC 1928 Rep); Endon (1935).

Briza media L.   Quaking-grass
Dry limestone pastures, calcareous sands, wet heathy roadsides, water meadows on rich soils and sometimes in sphagnum swamps: particularly common in the limestone districts of the north: B–P, R–T, W–a, d–g, i–n, r–v, z, 157. Pitt, 1794. (Map p 243).

Melica uniflora Retz.   Wood Melick
Woods and shady hedgebanks, particularly in Needwood Forest, on the northern limestone and in the south west: B, E–F, H–L, N–P, R, (S), T–U, W–X, Z–b, d, f–g, j, n, r–s, v–w, (x), y–z, 84. 'Weston-under-Lizard, in the Park' (Dickenson 1798). (Map p 243).

M. nutans L.   Mountain Melick
Rocky limestone woods: rare. Musden Wood (1944!); Apes Tor,
Hurt's Wood, Dovedale Wood (Shimwell 1968).

Bromus erectus Huds.   Upright Brome
Observed on the Silurian limestone at Dudley, Sedgley and Walsall,
and as a casual by the canals in a few places: M, W, d, m, s, 6.
Wren's Nest (Curtis and Druce in BEC 1923 Rep).

B. ramosus Huds.   Hairy-brome
Woods and shady hedgebanks, particularly on the Keuper marl
and limestone: B, G–x, z, 241. Dickenson, 1798. (Map p 244).

B. sterilis L.   Barren Brome
Roadsides and waste places: abundant in the south: B, E–z, 445.
Pitt, 1794.

B. mollis L.   Soft-brome
Meadows, hayfields, roadsides and waste places: B–C, E–z, 489.
Pitt, 1794.

B. thominii Hardouin   Lesser Soft-brome
Cornfields and roadsides: probably common, but overlooked.
Tyrley (1948!); cornfield near Loynton Moss (1948!); roadside
near Wolseley Bridge (1948!); Stowe (1956!); Pipe Ridware
(1956!). First three det C. E. Hubbard.

B. lepidus Holmberg   Slender Brome
Arable fields, sandy roadsides and waste ground: frequent: H, K,
Q–T, V–W, d–g, j, n, r, w, y, 24. Near Cresswell, Draycott-in-the-
Moors, det W. O. Howarth (1939!).

B. racemosus L.   Smo   oth Brome
Waterhouses, det Hubbard (1945!).

B. commutatus Schrad.   Meadow Brome
Burton (Druce in BEC 1930 Rep). There are earlier records which
may be right.

B. arvensis L.   Field Brome
Burton (Druce and Curtis in BEC 1926 Rep). Earlier records
uncertain.

B. secalinus L.   Rye Brome
Burton (Curtis 1930). Earlier records uncertain.

Brachypodium sylvaticum (Huds.) Beauv.   False-brome
Woods and shady roadside banks, especially on the limestone and
Keuper marl: B, H–L, N–b, d–z, 209. Garner, 1844. (Map p 244).

B. pinnatum (L.) Beauv.   Tor-grass
Casual. Cheadle Road, Oakamoor (Berrisford 1908!); by the canal
at Wombourn and Tettenhall (1954!).

Agropyron caninum (L.) Beauv.   Bearded Couch
Wood margins and shady hedgebanks, chiefly on the eastern side
of the county: H–L, (M), N–P, S–V, X–b, d, (f), g–k, n–r, t–w, z, 81.
Garner, 1844. (Map p 244).

A. repens (L.) Beauv.   Common Couch
Roadsides, arable fields, waste ground: B–z, 679. Pitt, 1794.

Hordeum secalinum Schreb.   Meadow Barley
Meadows, chiefly in the Trent basin and along the lower reaches
of the Dove: local: (N), S, (T–U), V, X–Y, (Z), a–b, (g), h, p, w,
y, 14. Pitt, 1794.

H. murinum L.   Wall Barley
Walls, roadsides and waste ground, particularly in the town areas
of south Staffordshire: abundant in and about Stafford, Burton,
Lichfield and in the Black Country: J–L, (N), U, X–b, d–y, 219.
'Old walls and bridges' (Dickenson 1798). The pattern was the same
a hundred years ago: 'Waste places, but rare in the north of the
county; Lichfield Close; near Stourbridge' (Garner 1844).
(Map p 244).

Hordelymus europaeus (L.) Harz   Wood Barley
Woods on the Carboniferous limestone: rare. Hurt's Wood,
Dovedale Wood, Cheshire Wood (Shimwell 1968).

Koeleria cristata (L.) Pers.   Crested Hair-grass
Restricted to dry hillsides on the Carboniferous limestone, but
there common: H–J, N–P, 14. Garner, 1844.

Trisetum flavescens (L.) Beauv.   Yellow Oat-grass
Pastures and dry road verges: B–z, 377. Dickenson, 1798.

Avena fatua L.   Wild-oat
'A weed in cornfields' (Pitt 1817); Codsall (Fraser 1879!); Trysull
(Fraser 1882!); Rolleston (1956!); Hollington (1960!).

A. sativa L. Oat
Often found in waste places as a relic of cultivation. Stafford (Moore 1889!).

A. strigosa Schreb. Bristle Oat
Cornfields, Burton (Brown 1863); Trysull (Fraser 1879!).

Helictotrichon pratense (L.) Pilg. Meadow Oat-grass
Disused quarries and hillsides on the Carboniferous limestone: frequent: H–J, N, (b), (f), (t), 6. Garner, 1844. Unconfirmed records for Burton (Flora 1901); canal at Armitage and Hay Head lime works (Bagnall 1901).

H. pubescens (Huds.) Pilg. Downy Oat-grass
Pastures and roadsides, chiefly on the limestone, but more widely distributed than *H. pratense* and much commoner: C, G–K, N–P, S–U, Y–Z, b, d, (f), g–h, m–p, t, (v), 61. Garner, 1844. (Map p 245).

Arrhenatherum elatius (L.) Beauv. ex J. and C. Presl False Oat-grass
Waysides and waste places: A–z, 771. Pitt, 1794. 'In every hedge and cornfield, root curiously knotted and troublesome to the farmer' (Garner 1844).

Holcus lanatus L. Yorkshire-fog
Fields, roadsides, waste ground: A–z, 788. Pitt, 1794.

H. mollis L. Creeping Soft-grass
In many habitats, but particularly common in bluebell woods: A–z, 637. Pitt, 1794. 'Cornfields and hedges, the root is a very rank couch and extremely troublesome in arable lands' (Dickenson 1798).

Deschampsia cespitosa (L.) Beauv. Tufted Hair-grass
Undrained fields and wet places throughout the county: A–u, w–z, 655. Pitt, 1794.

D. flexuosa (L.) Trin. Wavy Hair-grass
Dry acid heaths: locally dominant on the Millstone grit and Bunter sandstone: B–U, W–g, j–n, r–z, 369. Dickenson, 1798.

Aira praecox L. Early Hair-grass
Locally plentiful in open places in dry shallow sandy soil, less common about outcrops of rock on the Carboniferous limestone: C, G–P, R–S, U, X–Z, (a), c, (d), e–g, k–n, r, (v), w–y, 72. 'Gravel pit beyond Woodmill Brook, 1791' (Gisborne MS). 'Heaths,

Weston-under-Lizard, on hedge banks formerly common land' (Dickenson 1798).

A. caryophyllea L. Silver Hair-grass
In similar situations to the last species and often growing with it, but more common on the limestone: J–K, P, R–S, W, Y–Z, (d), e–g, j–m, r, t, (v), w, 34. 'With *Aira praecox* and by Lush Pool, 1791' (Gisborne MS). 'Blymhill, in light sandy land' (Dickenson 1798).

Calamagrostis epigejos (L.) Roth Wood Small-reed
Wet places by roadsides and in woods: rare. 'Blymhill, in the Watling-street way, in a hedge between Ivetsy-bank and Stinking-lake; on the border of Aqualate Mere, on the south side, near the boathouse' (Dickenson 1798); Callingwood (Brown 1863); hedge between Codsall and Codsall Wood (Fraser 1879!); Kingston Pool (Bagnall 1901); roadside between Fullmoor Wood and Hatherton (1944!); Bilston (1951!); Hamps Valley, 'a small patch on a bushy grassy slope only a few yards from the old railway track' between Waterhouses and Beeston Tor, Miss Hollick (1953); Teddesley Hay, on the northern boundary of the park (1961!).

C. canescens (Weber) Roth Purple Small-reed
Wet peaty woods: local. Aqualate Mere (Stokes in Withering 1796, Fraser 1864! Edees 1953!); Pensnett Reservoir (Scott 1832); Kingston Pool (Bagnall 1897!) and Kingston Brook (1959!); Chartley Moss (1944!); Shelmore Wood (1949!); Coneygreave Haft near Norbury (1953!); Walton's Wood.

Agrostis canina L. Brown Bent
Acid heaths and woods in both dry and wet soils: probably common, but incompletely recorded: B–C, H, K, N, R, T, Z–a, (e), f, m, (t), w–x, z, 19. Dickenson, 1798.

A. tenuis Sibth. Common Bent
Pastures and roadsides: A–z, 749. Pitt, 1794.

A. gigantea Roth Black Bent
Sandy arable fields: frequent: M, (T), U, X, Z, e, (f), k–m, (r), s, (t), v–w, y, 19. 'In cold clayey arable land' (Dickenson 1798).

A. stolonifera L. Creeping Bent
Meadows, damp roadsides, arable and waste land: B–D, G–z, 431. Pitt, 1794.

Phleum bertolonii DC.   Smaller Catstail
Known to be common in dry soils, but records have not been kept.
Dovedale (Smith 1871).

P. pratense L.   Timothy
Common in water meadows and hayfields as a native and cultivated
plant, and as an aggregate species, including *P. bertolonii*, distributed as follows: A–z, 661. Pitt, 1794.

Alopecurus myosuroides Huds.   Black-grass
Locally common in cultivated fields in former days, but now rare
and recorded recently only as a casual: (N), (Y–Z), b, (f–g), j, p,
(t), 3. 'Cornfields between Hadley End and Miss Riley's Wood
abundantly, 1791' (Gisborne MS). 'A troublesome weed, rare, in
wet arable land at Mavesyn Ridware' (Dickenson 1798). 'Abundant
about Stone and Stafford' (Garner 1844). Corroborating specimens
from the vicinity of Stafford (Fraser 1864! and Moore 1889!).
Recent records for Burton (1944!); Tamworth, canal at Bolehall
(1958); Brewood, a single plant in a reseeded field (1962).

A. pratensis L.   Meadow Foxtail
Fields and roadsides: A–z, 725. Pitt, 1794.

A. geniculatus L.   Marsh Foxtail
'Wet muddy places' (Brown 1863): A–u, w–y, 517. Blymhill
(Dickenson 1798).

A. aequalis Sobol.   Orange Foxtail
In mud at the edges of lakes and reservoirs when the water has
receded in a dry summer: rare, but sometimes locally abundant:
B, (F), G, W, d–e, j–k, (m), w, 8. 'Near the railway station, Burton'
(Brown in Garner 1844). Recent records for Rudyard Lake (1938!);
Gailey Reservoir (1947!); Foucher's Pool (1954!); Knighton
Reservoir (1954!); Chillington Pool, SJ/8505 (1954!); Calf Heath
Wood (1954!); Deep Hayes Reservoir in 1955 (1956); pond at
Lower Mitton, SJ/886150 (1961!).

Milium effusum L.   Wood Millet
Damp woods: frequent: B, J–P, R–U, X, Z–d, f, j–n, r–s, (t), v–y,
72. 'Miss Riley's Wood, 1791' (Gisborne MS). (Map p 245).

Anthoxanthum odoratum L.   Sweet Vernal-grass
'In meadows and pastures, frequent, as well in wet, marshy, as dry,

upland soils' (Dickenson 1798): A–z, 742. 'Very common all over Needwood Forest' (Pitt 1794).

A. puelii Lecoq and Lamotte   Annual Vernal-grass
Whittington near Kinver (Fraser 1877!); sandy ground, Hawkesyard (Reader 1900!); Burton (Druce in hb Wedgwood 1929! conf Hubbard).

Phalaris arundinacea L.   Reed Canary-grass
'Banks of rivers and moist ditch-banks, common' (Dickenson 1798): B–z, 384. (Map p 245).

P. canariensis L.   Canary-grass
A frequent bird-seed alien and known for many years as a casual on waste ground: C, F–H, K–N, R–U, Y–Z, b, e–f, k–m, r–s, (t), x–y, 42. 'Frequent but not wild' (Garner 1844).

Nardus stricta L.   Mat-grass
Heaths and rough pastures in acid soils: A–D, F–P, R–U, W–a, d–g, j–n, r–t, w–z, 232. 'The staple grass of some of our waste lands; on Cannock Heath' (Pitt 1796). (Map p 245).

DISTRIBUTION MAPS

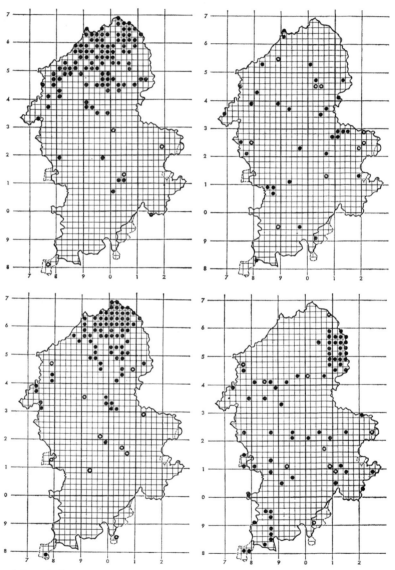

(*above left*) Equisetum sylvaticum   (*above right*) Equisetum telmateia
(*below left*) Blechnum spicant   (*below right*) Asplenium ruta-muraria

196

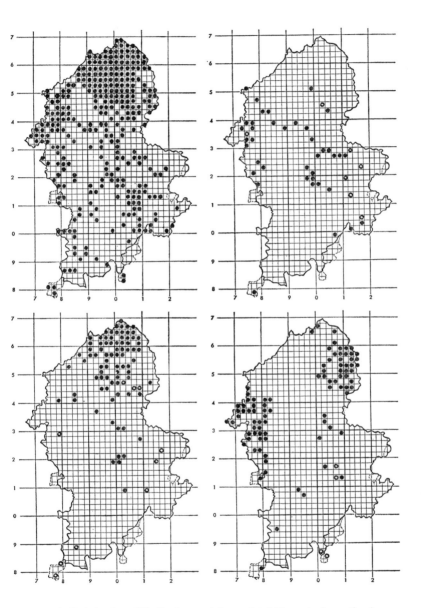

(*above left*) Athyrium filix-femina   (*above right*) Dryopteris carthusiana
(*below left*) Thelypteris limbosperma   (*below right*) Polypodium vulgare

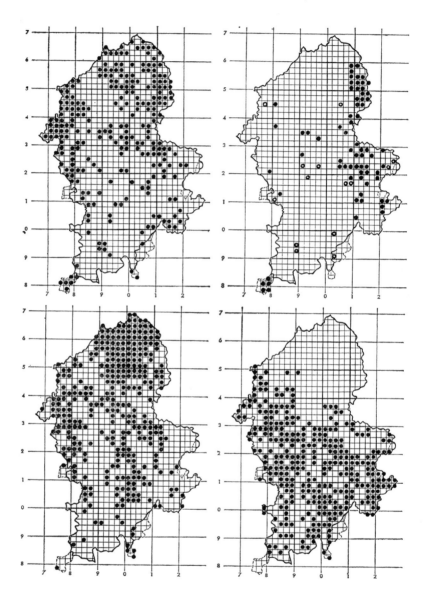

(*above left*) Anemone nemorosa   (*above right*) Ranunculus auricomus
(*below left*) Ranunculus flammula   (*below right*) Ranunculus sceleratus

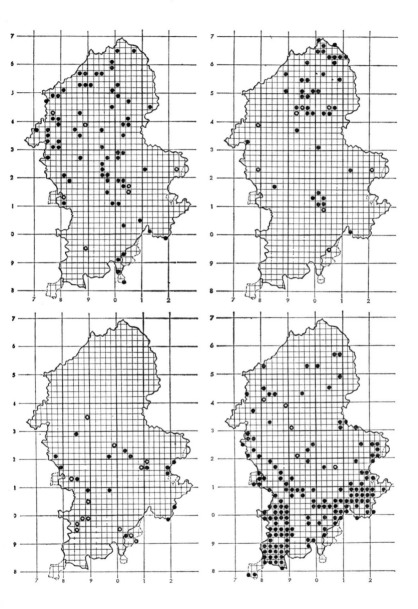

(*above left*) Ranunculus hederaceus (*above right*) Ranunculus omiophyllus
(*below left*) Thalictrum flavum   (*below right*) Papaver rhoeas

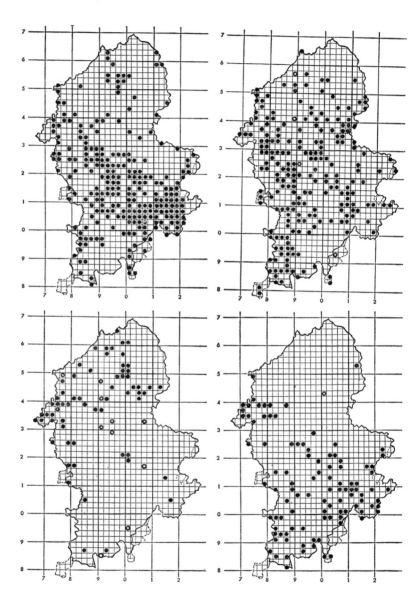

(*above left*) Papaver dubium
(*below left*) Corydalis claviculata

(*above right*) Chelidonium majus
(*below right*) Fumaria officinalis

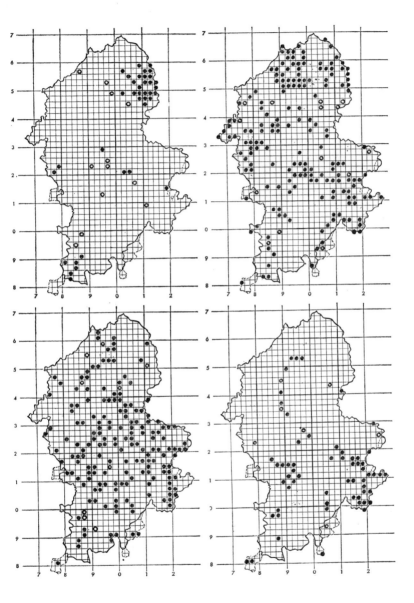

(*above left*) Erophila verna
(*below left*) Rorippa islandica

(*above right*) Cardamine amara
(*below right*) Rorippa amphibia

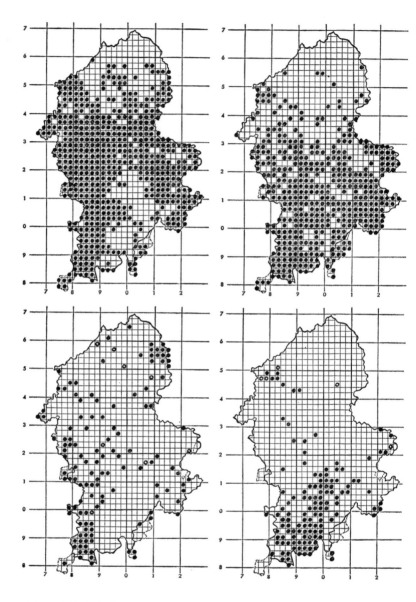

(*above left*) Alliaria petiolata
(*below left*) Arabidopsis thaliana

(*above right*) Sisymbrium officinale
(*below right*) Reseda luteola

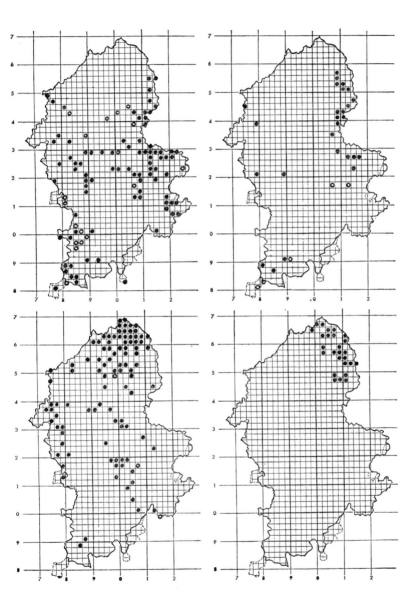

(*above left*) Viola odorata
(*below left*) Viola palustris

(*above right*) Viola reichenbachiana
(*below right*) Viola lutea

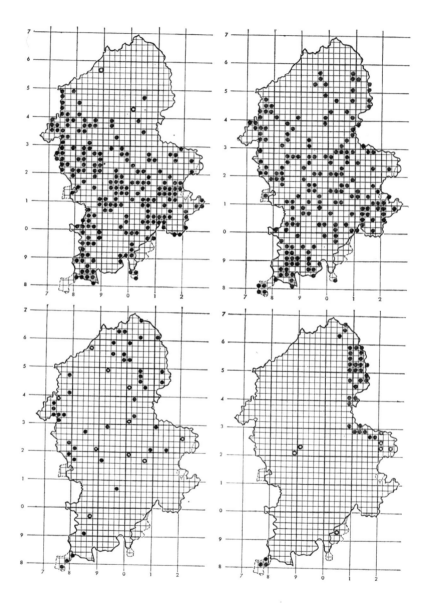

(*above left*) Viola arvensis
(*below left*) Hypericum pulchrum

(*above right*) Hypericum perforatum
(*below right*) Hypericum hirsutum

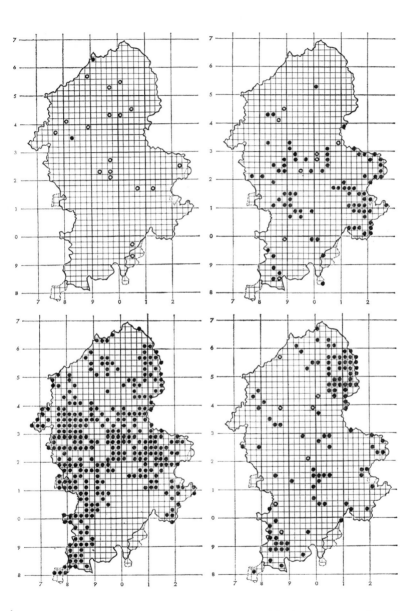

(*above left*) Agrostemma githago (*above right*) Myosoton aquaticum
(*below left*) Moehringia trinervia (*below right*) Arenaria serpyllifolia

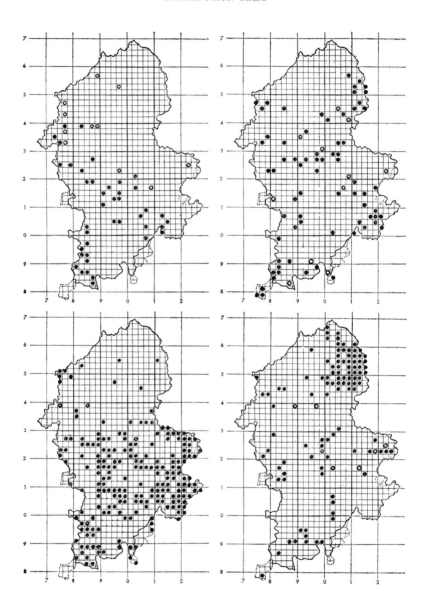

(*above left*) Scleranthus annuus
(*below left*) Malva sylvestris

(*above right*) Malva moschata
(*below right*) Linum catharticum

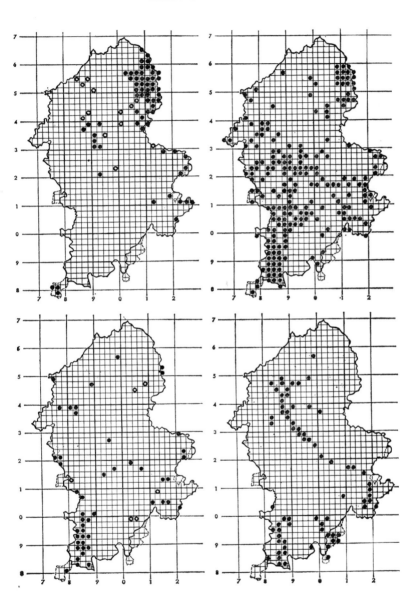

(*above left*) Geranium pratense
(*below left*) Erodium cicutarium

(*above right*) Geranium molle
(*below right*) Impatiens glandulifera

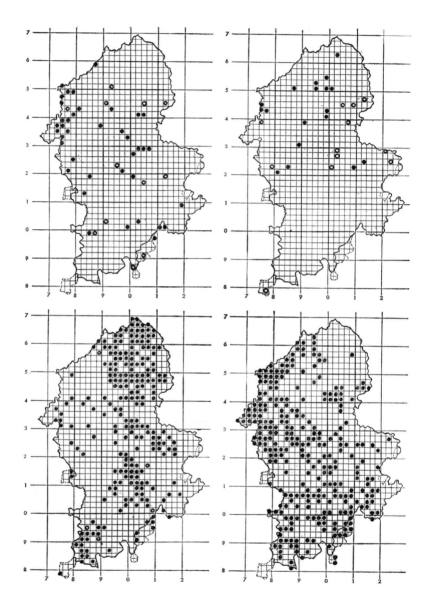

(*above left*) Frangula alnus
(*below left*) Ulex gallii

(*above right*) Genista tinctoria
(*below right*) Sarothamnus scoparius

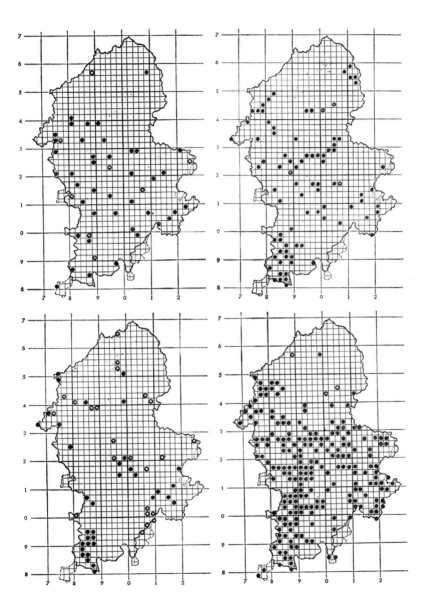

(*above left*) Ononis repens
(*below left*) Ornithopus perpusillus

(*above right*) Trifolium campestre
(*below right*) Vicia hirsuta

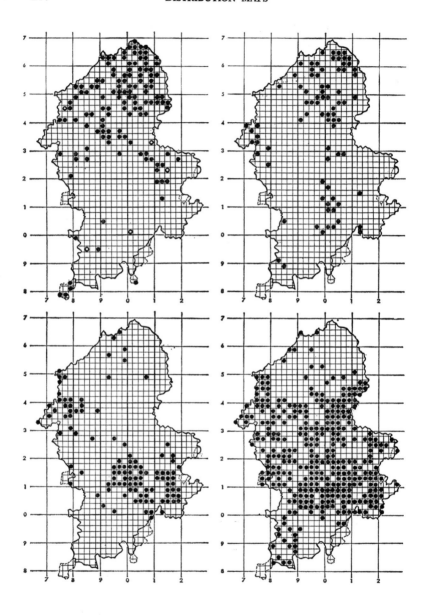

(*above left*) Lathyrus montanus
(*below left*) Rubus selmeri

(*above right*) Rubus scissus
(*below right*) Rubus lindleianus

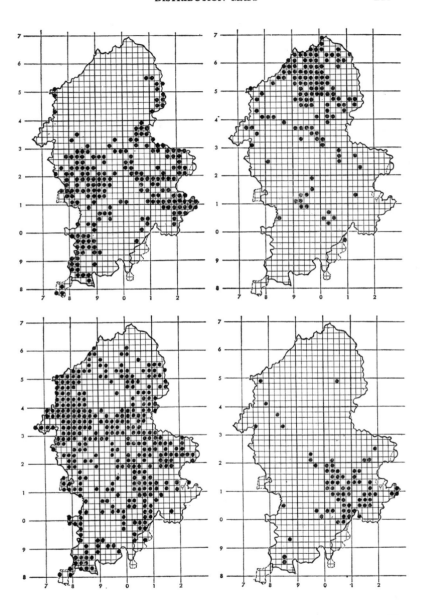

(*above left*) Rubus ulmifolius
(*below left*) Rubus vestitus

(*above right*) Rubus sprengelii
(*below right*) Rubus leightonii

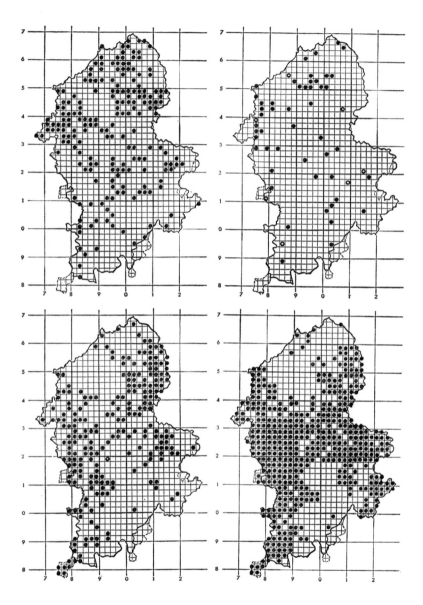

(*above left*) Rubus dasyphyllus
(*below left*) Fragaria vesca

(*above right*) Potentilla palustris
(*below right*) Geum urbanum

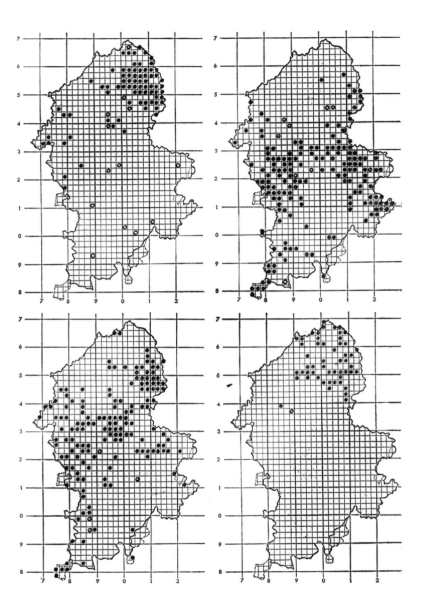

(*above left*) Geum rivale   (*above right*) Agrimonia eupatoria
(*below left*) Alchemilla vestita   (*below right*) Alchemilla xanthochlora

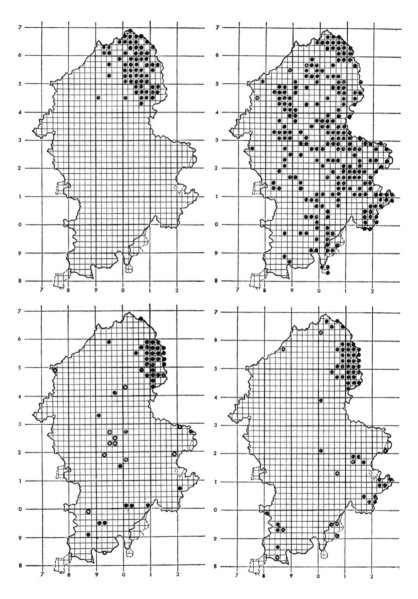

(*above left*) Alchemilla glabra
(*below left*) Sedum acre

(*above right*) Sanguisorba officinalis
(*below right*) Saxifraga granulata

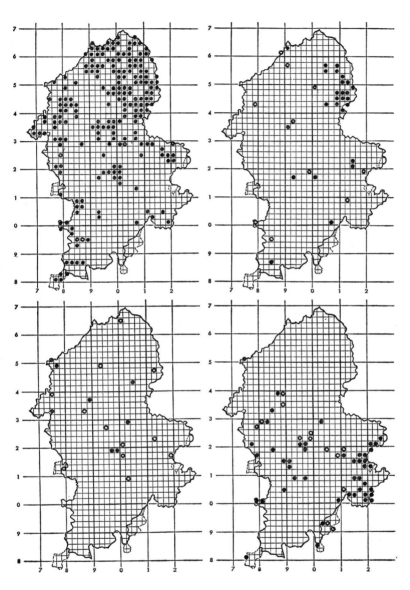

(*above left*) Chrysosplenium
oppositifolium
(*below left*) Drosera rotundifolia

(*above right*) Chrysosplenium
alternifolium
(*below right*) Lythrum salicaria

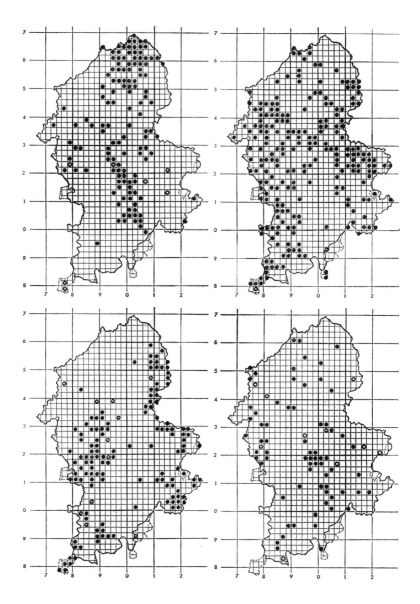

(*above left*) Epilobium palustre
(*below left*) Swida sanguinea

(*above right*) Circaea lutetiana
(*below right*) Hydrocotyle vulgaris

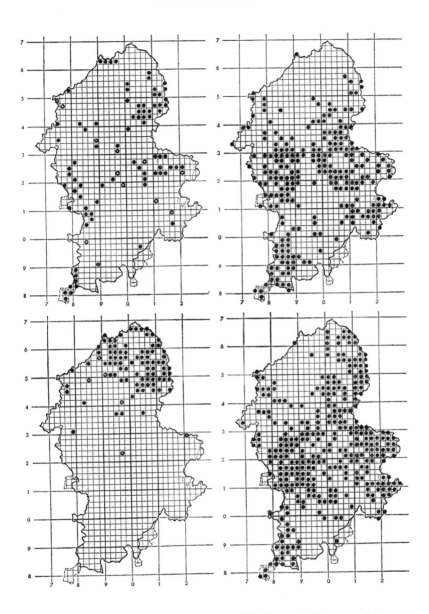

(*above left*) Sanicula europaea (*above right*) Chaerophyllum temulentum
(*below left*) Myrrhis odorata   (*below right*) Torilis japonica

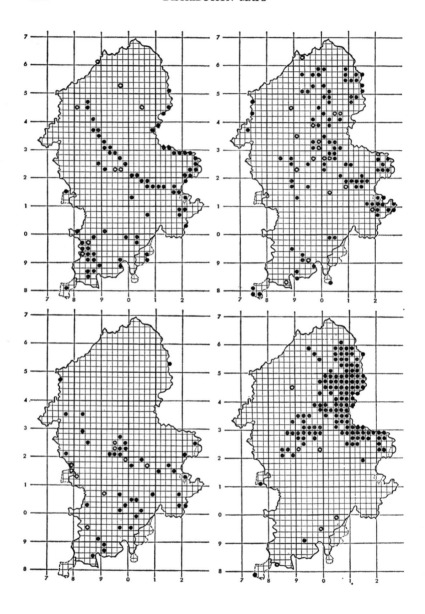

(*above left*) Conium maculatum   (*above right*) Pimpinella saxifraga
(*below left*) Berula erecta   (*below right*) Pimpinella major

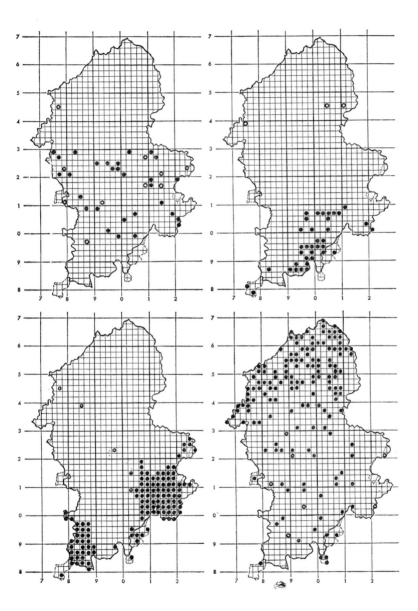

(*above left*) Oenanthe fistulosa (*above right*) Oenanthe crocata
(*below left*) Bryonia dioica (*below right*) Polygonum bistorta

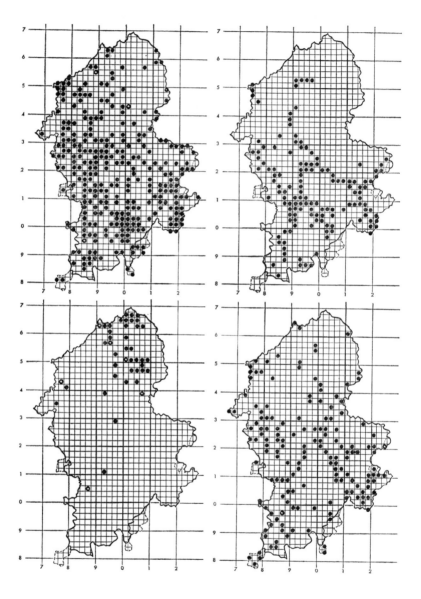

(*above left*) Polygonum amphibium (*above right*) Rumex hydrolapathum
(*below left*) Rumex alpinus (*below right*) Humulus lupulus

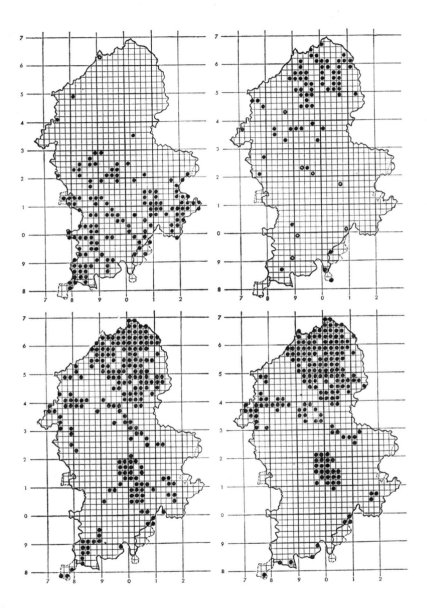

(*above left*) Ulmus procera
(*below left*) Calluna vulgaris

(*above right*) Salix pentandra
(*below right*) Vaccinium myrtillus

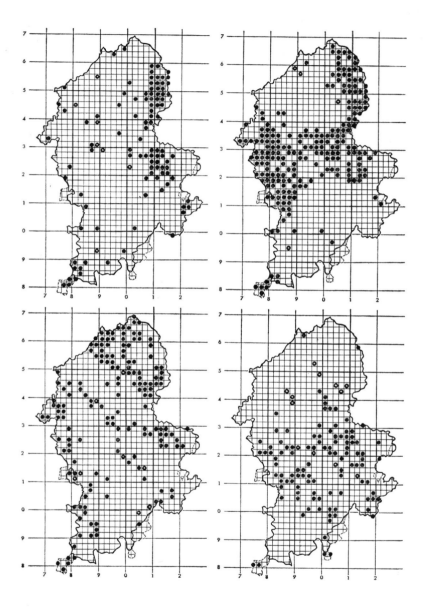

(*above left*) Primula veris   (*above right*) Primula vulgaris
(*below left*) Lysimachia nemorum   (*below right*) Lysimachia nummularia

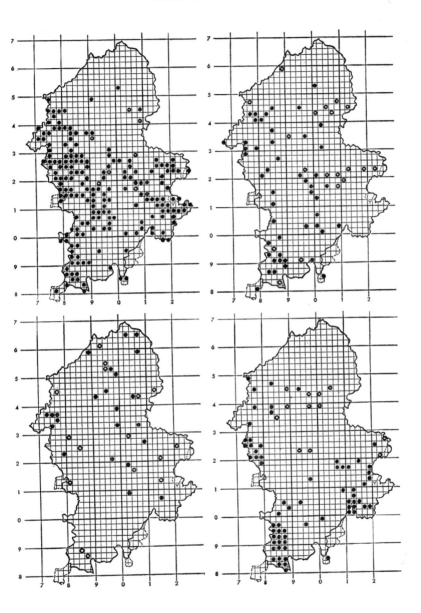

(*above left*) Anagallis arvensis
(*below left*) Menyanthes trifoliata

(*above right*) Centaurium erythraea
(*below right* (Lycopsis arvensis

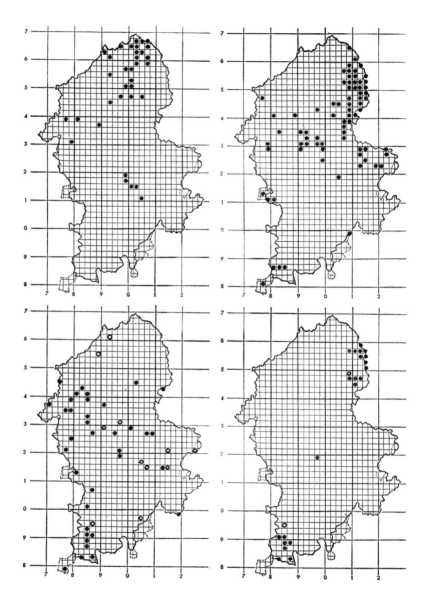

(*above left*) Myosotis secunda
(*below left*) Myosotis discolor

(*above right*) Myosotis sylvatica
(*below right*) Myosotis ramosissima

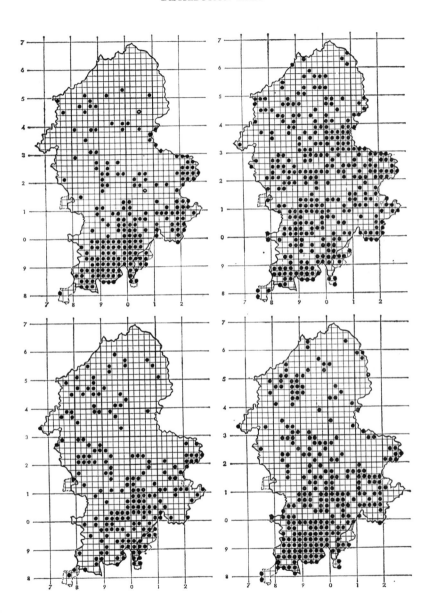

(*above left*) Convolvulus arvensis
(*above left*) Calystegia silvatica

(*above right*) Calystegia sepium
(*below right*) Linaria vulgaris

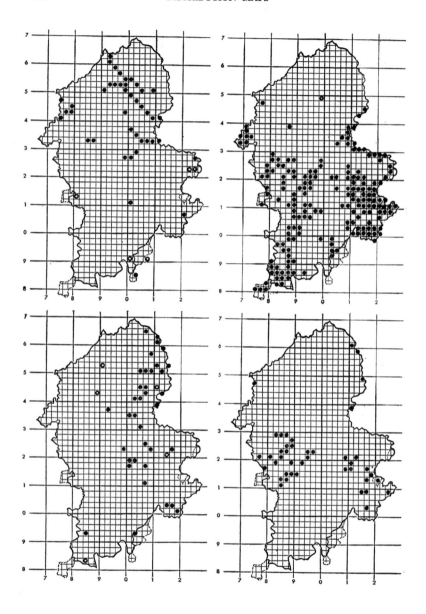

(*above left*) Chaenorhinum minus   (*above right*) Scrophularia auriculata
(*below left*) Mimulus guttatus   (*below right*) Veronica catenata

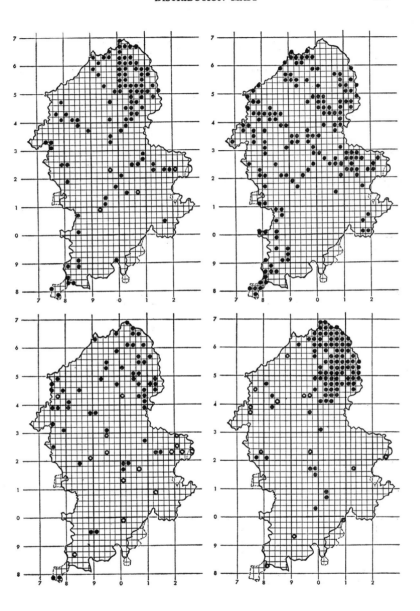

(*above left*) Veronica officinalis
(*below left*) Pedicularis sylvatica

(*above right*) Veronica montana
(*below right*) Euphrasia officinalis

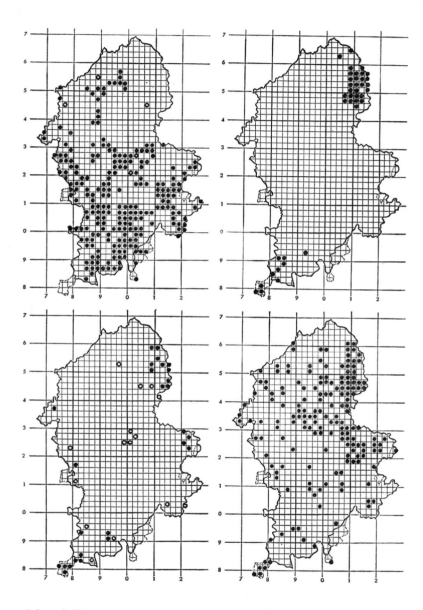

(*above left*) Lycopus europaeus
(*below left*) Clinopodium vulgare

(*above right*) Thymus drucei
(*below right*) Betonica officinalis

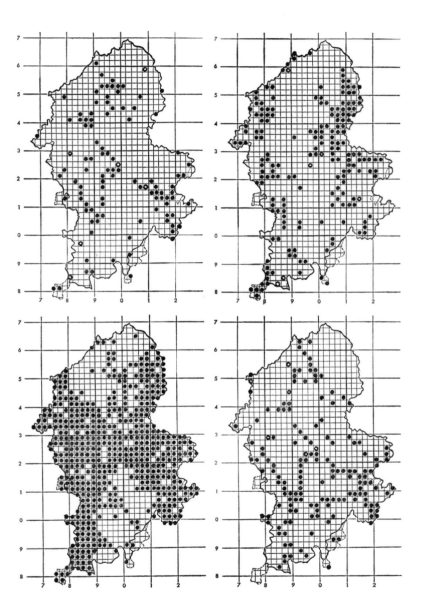

(*above left*) Stachys palustris   (*above right*) Lamiastrum galeobdolon
(*below left*) Glechoma hederacea   (*below right*) Scutellaria galericulata

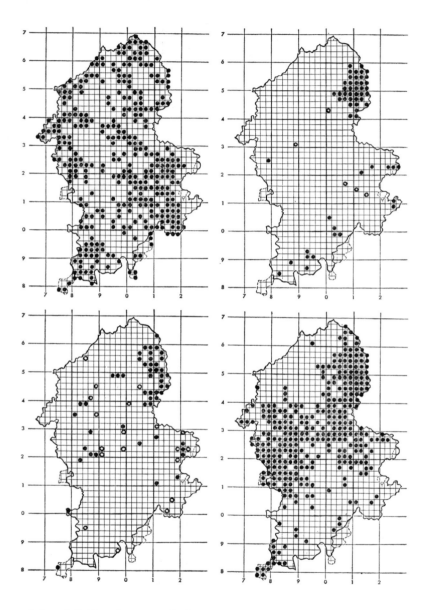

(*above left*) Teucrium scorodonia
(*below left*) Campanula latifolia

(*above right*) Plantago media
(*below right*) Cruciata laevipes

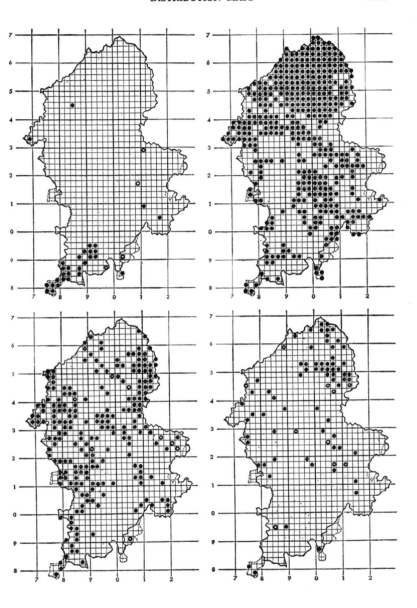

(*above left*) Galium mollugo
(*below left*) Adoxa moschatellina

(*above right*) Galium saxatile
(*below right*) Valeriana dioica

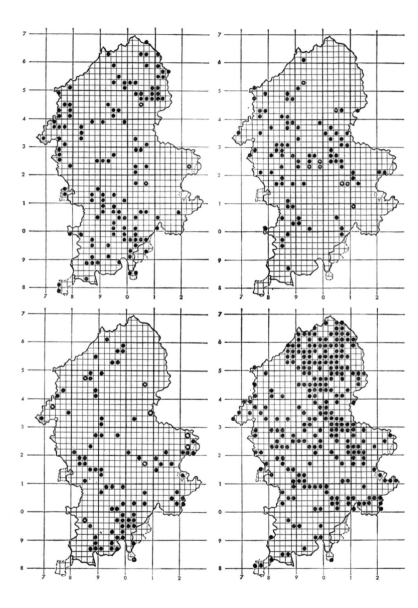

(*above left*) Knautia arvensis
(*below left*) Bidens tripartita

(*above right*) Bidens cernua
(*below right*) Senecio aquaticus

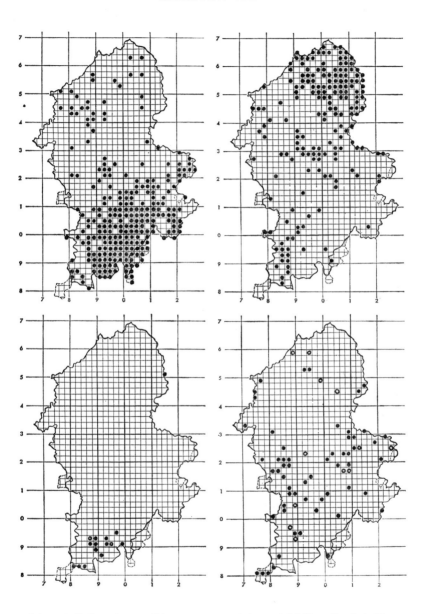

(*above left*) Senecio squalidus
(*below left*) Inula conyza

(*above right*) Petasites hybridus
(*below right*) Pulicaria dysenterica

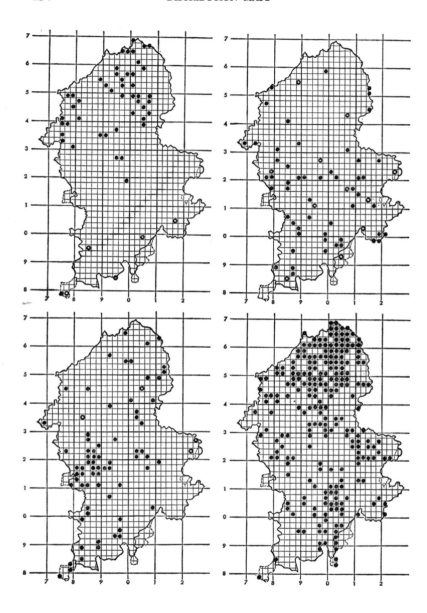

(*above left*) Solidago virgaurea   (*above right*) Eupatorium cannabinum
(*below left*) Anthemis cotula   (*below right*) Achillea ptarmica

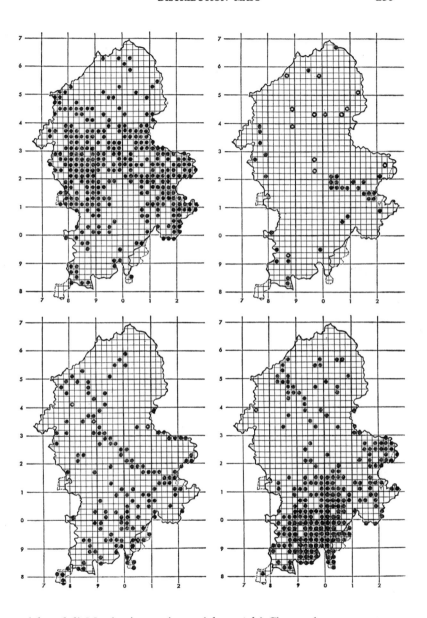

(*above left*) Matricaria recutita (*above right*) Chrysanthemum segetum
(*below left*) Tanacetum vulgare (*below right*) Artemisia absinthium

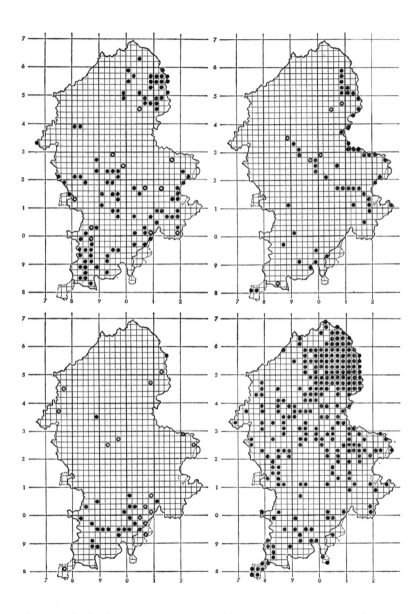

(*above left*) Carduus nutans   (*above right*) Carduus acanthoides
(*below left*) Centaurea scabiosa   (*below right*) Leontodon hispidus

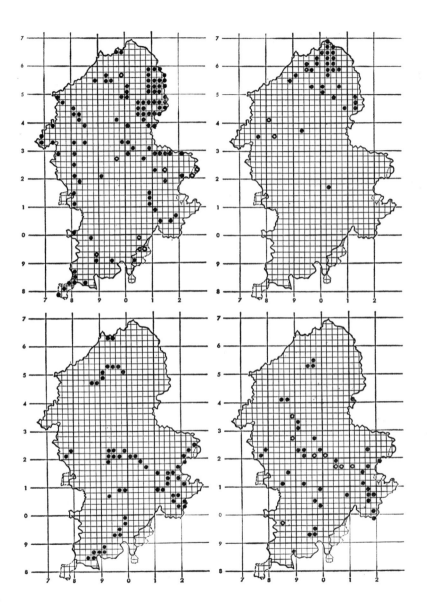

(*above left*) Mycelis muralis
(*below left*) Sagittaria sagittifolia

(*above right*) Crepis paludosa
(*below right*) Butomus umbellatus

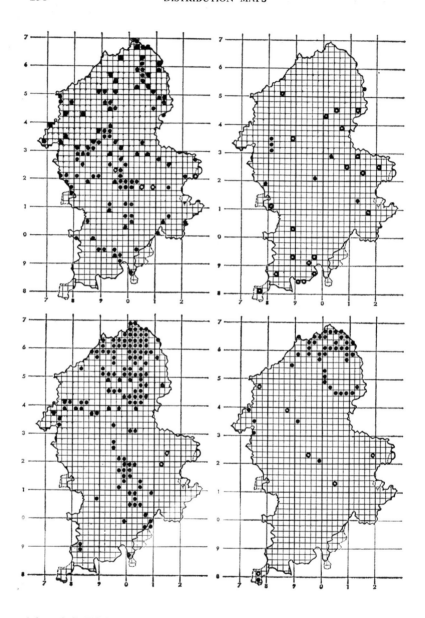

(*above left*) Triglochin palustris
(*below left*) Juncus squarrosus

(*above right*) Paris quadrifolia
(*below right*) Luzula sylvatica

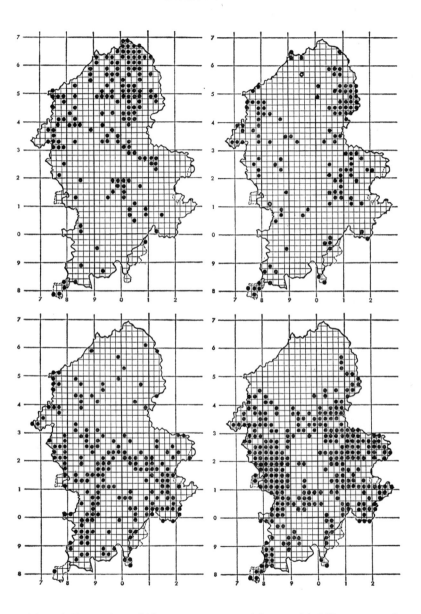

(*above left*) Luzula multiflora
(*below left*) Iris pseudacorus

(*above right*) Allium ursnumi
(*below right*) Tamus communis

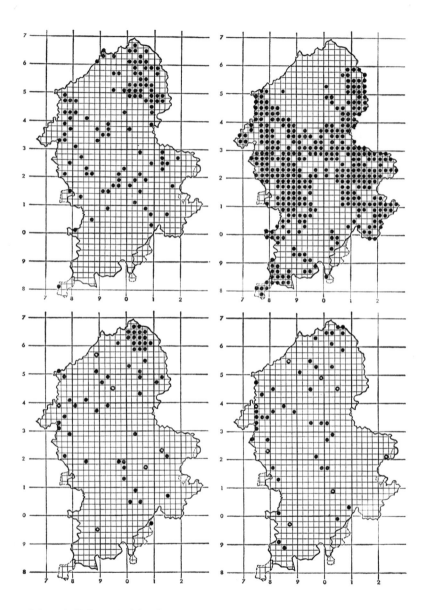

(*above left*) Dactylorhiza fuchsii
(*below left*) Eriophorum angustifolium

(*above right*) Arum maculatum
(*below right*) Carex rostrata

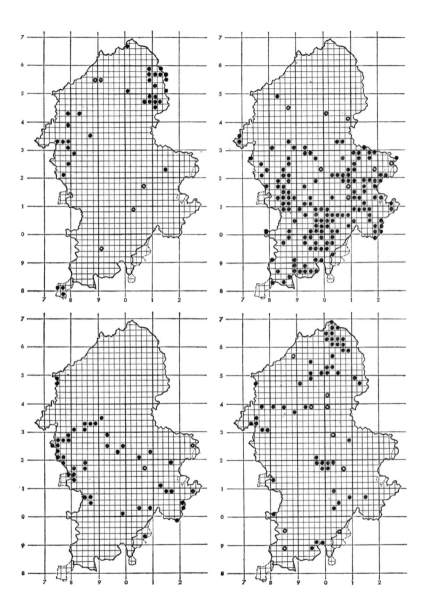

(*above left*) Carex caryophyllea   (*above right*) Carex otrubae
(*below left*) Carex disticha   (*below right*) Carex echinata

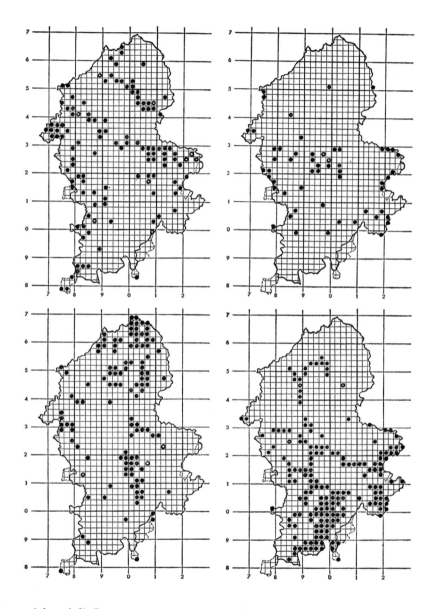

(*above left*) Carex remota
(*below left*) Molinia caerulea

(*above right*) Phragmites australis
(*below right*) Glyceria maxima

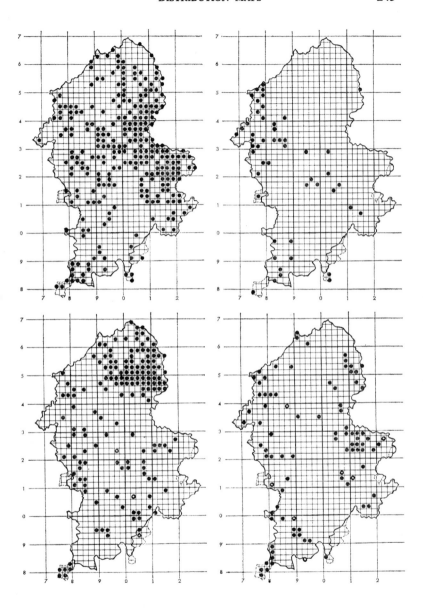

(*above left*) Festuca gigantea
(*below left*) Briza media

(*above right*) Vulpia bromoides
(*below right*) Melica uniflora

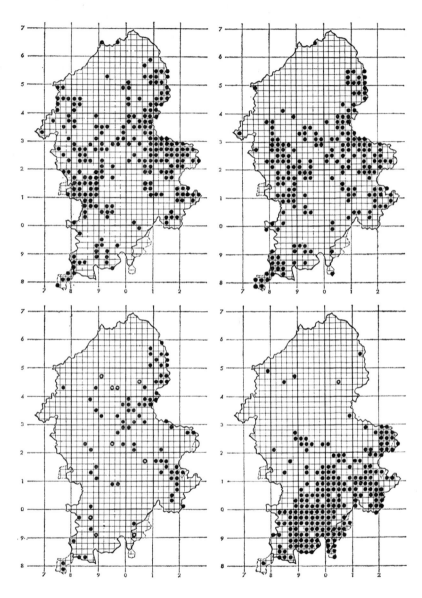

(*above left*) Bromus ramosus   (*above right*) Brachypodium sylvaticum
(*below left*) Agropyron caninum   (*below right*) Hordeum murinum

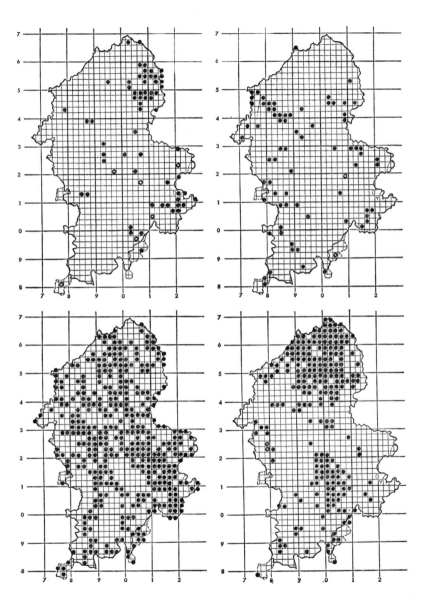

(*above left*) Helictotrichon pubescens (*above right*) Milium effusum
(*below left*) Phalaris arundinacea (*below right*) Nardus stricta

GAZETTEER WITH NATIONAL GRID REFERENCES

Abbots Bromley	SK 0824	Barr Common	SP 0796
Abraham's Valley	SK 0020	Barr Farm	SP 0796
Acton	SJ 9318	Barrow Hill	SO 9189
Acton Hill	SJ 9419	Barrowhill (Rocester)	SK 1140
Acton Mill Bridge	SJ 9318	Barton	SK 1818
Adbaston	SJ 7627	Barton Mill	SK 2016
Admaston	SK 0523	Barton Turn	SK 2018
Aldridge	SK 0500	Basford	SJ 9851
Aldridge Wharf Bridge	SK 0401	Baswich	SJ 9621
Allimore Green Common	SJ 8519	Bearsbrook	SK 0133
Alrewas	SK 1615	Beaudesert Old Park	SK 0413
Alstonfield	SK 1355	Beech	SJ 8538
Alton	SK 0742	Beech House	SO 8493
Alton Towers	SK 0743	Beeston Tor	SK 1054
Amblecote	SO 9085	Beffcote	SJ 8019
Anslow	SK 2125	Bellamour House	SK 0420
Apes Tor	SK 1058	Belmont	SK 0049
Aqualate Mere	SJ 7720	Belvide Reservoir	SJ 8510
Aquamoor	SJ 8314	Bentley Farm	SK 0818
Archford Moor	SK 1158	Beresford Dale	SK 1259
Arley	SO 7680	Berry Ring	SJ 8821
Arley Wood	SO 8082	Betley	SJ 7548
Armitage	SK 0716	Betley Common	SJ 7448
Ashley	SJ 7636	Betley Mere	SJ 7448
Ashwood	SO 8688	Biddulph	SJ 8857
Aston Cliff	SJ 7642	Biddulph Castle	SJ 8960
Audley	SJ 7950	Biddulph Grange	SJ 8959
Axe Edge	SK 0369	Big Hyde Rough	SJ 8608
		Big Peg's Wood	SK 1549
Back of Ecton	SK 1057	Bilbrook	SJ 8703
Baddeley Edge	SJ 9151	Bilston	SO 9596
Baggeridge Wood	SO 8993	Bincliff Thicket	SK 1153
Bagnall	SJ 9250	Birches Valley	SK 0116
Bagot's Park	SK 0927	Birchills	SO 9999
Balterley Heath	SJ 7450	Birchwood Park	SK 0033
Banktop Wood	SK 1328	Bishop's Wood	SJ 7531
Bar Hill	SJ 7643	Bishop's Wood (Brewood)	SJ 8409
Bar Lane	SK 1718	Bishton	SK 0220
Barlaston	SJ 8938	Black Bank	SJ 8147
Barlaston Common	SJ 9239	Black Heath	SK 0549
Barr Beacon	SP 0697	Black Lake	SJ 8539

Blacklake Farm	SJ 9338	Bushbury	SJ 9202
Black Mere	SJ 7450	Butter Bank	SJ 8723
Blakeley	SJ 9747	Buttermilk Hill	SK 1028
Blithbury	SK 0820	Butterton (Leek)	SK 0756
Blithfield	SK 0423	Butterton (Newcastle)	SJ 8342
Blithfield Reservoir	SK 0525	Butterton Park	SJ 8342
Blore	SK 1349	Byrkley Lodge	SK 1623
Blymhill	SJ 8012		
Blymhill Lawn	SJ 8211	Caldonlow	SK 0748
Bobbington	SO 8090	Calf Heath	SJ 9209
Bolehall	SK 2104	Callingwood	SK 1823
Borough End	SK 2117	Callingwood Lane	SK 1923
Boscobel House	SJ 8308	Calton	SK 1050
Bradeshall Farm	SO 9790	Caltonmoor	SK 1148
Bradley	SJ 8718	Calwich Abbey and Park	SK 1243
Bradnop	SK 0154	Camp Farm	SO 8489
Bradwell	SJ 8450	Camp Hills	SJ 7740
Brakenhurst Bog	SK 1423	Cannock Chase	SJ 9–1–
Brancote Covert	SJ 9621	Cannock Chase Reservoir	SK 0307
Brand Plantation	SK 0466	Castern	SK 1252
Branston	SK 2221	Castlecroft	SO 8797
Brereton	SK 0516	Catholme	SK 1916
Brereton Colliery	SK 0415	Cauldon	SK 0749
Brereton Cross	SK 0615	Caverswall	SJ 9542
Brewood	SJ 8808	Cawarden Springs	SK 0618
Bridestones	SJ 9062	Chapman's Wood	SJ 7916
Brierley	SO 9186	Chartley Moss	SK 0228
Brindley Heath	SK 0115	Chatcull	SJ 7934
Brindley Valley	SJ 9915	Cheadle	SK 0043
Brinsford	SJ 9205	Chebsey	SJ 8628
Brittle's Farm	SO 7982	Checkley	SK 0237
Broad Heath	SJ 8525	Cheddleton	SJ 9752
Broadhurst Green	SJ 9815	Cheddleton Heath	SJ 9853
Brockhurst Coppice	SJ 8212	Cheshire Wood	SK 1153
Brocton	SJ 9619	Chesterfield	SK 1005
Brook House Farm	SK 1130	Chillington Pool	SJ 8505
Brookside Farm	SK 1130	Chorley	SK 0711
Brown Edge	SJ 9053	Chorlton Moss	SJ 7939
Brown Edge (Stanton)	SK 1246	Church Eaton	SJ 8417
Brownhills	SK 0505	Churnet Valley	SK 0–4–
Bucknall	SJ 9047	Clayhanger	SK 0404
Bunster Hill	SK 1451	Clayton	SJ 8543
Burlington	SJ 7711	Cliffe Park Station	SJ 9461
Burnt Wood	SJ 7335	Clifford's Wood Farm	SJ 8436
Burslem	SJ 8749	Cloud End House	SJ 9163
Burston	SJ 9330	Cloud Side	SJ 9063
Burton	SK 2423	Clough Head	SK 0859
Burton Abbey	SK 2522	Coalpit Slade	SK 1422

Coatestown	SK 0666	Dimsdale	SJ 8448
Cock Lane	SJ 9517	Dingle Cottage	SJ 8308
Codsall	SJ 8603	Doley Common	SJ 8121
Coldwall Bridge	SK 1449	Dove Bridge	SK 2328
Colton	SK 0520	Dovedale	SK 1–5–
Colwich	SK 0121	Dovedale Wood	SK 1452
Comberford	SK 1907	Doveridge Bridge	SK 1034
Compton	SO 8898	Downs Banks	SJ 8936
Coneygreave Haft	SJ 7923	Drabber Tor	SK 1356
Congreve	SJ 9013	Draycott in the Moors	SJ 9840
Consall	SJ 9748	Dudley and Dudley Castle	SO 9490
Consall Forge	SJ 9949	Dun Cow's Grove	SK 0466
Consall Old Hall	SJ 9849	Dunsley Dene	SO 8583
Consall Wood	SJ 9850	Dunstall	SK 1820
Cooksgate	SJ 7748	Dunstal Pool	SK 0826
Coombes Valley	SK 0052	Dunston	SJ 9217
Coppice Farm	SJ 9034	Dydon Wood	SK 1344
Coton Field	SJ 9224		
Coton in the Clay	SK 1629	Easthill Wood	SK 2021
Cotton Dell	SK 0545	Eastwall	SK 0344
Cotton Hall	SK 0646	Eccleshall	SJ 8329
County Lane	SJ 8303	Ecton	SK 0958
Cracow Moss	SJ 7447	Edingale	SK 2112
Craddocks Moss	SJ 7748	Elford	SK 1810
Cranmoor Wood	SJ 8400	Elford Station	SK 1910
Cresswell	SJ 9739	Ellastone	SK 1143
Creswell (Stafford)	SJ 9026	Ellenhall	SJ 8426
Croftbottom Farm	SK 0467	Elm Cottage	SJ 8708
Crowdecote	SK 1065	Elmhurst	SK 1112
Croxall	SK 1913	Endon	SJ 9253
Croxden	SK 0639	Enville	SO 8286
Croxden Abbey	SK 0639	Enville Common	SO 8386
Curborough Wood	SK 1212	Essington	SJ 9603
Dale Common	SJ 8721	Fairoak	SJ 7632
Dane Valley	SJ 9–6–	Farewell	SK 0811
Darlaston	SO 9897	Farley	SK 0644
Darlaston (Stone)	SJ 8835	Fauld	SK 1828
Darlaston Hall	SJ 8834	Fawfieldhead	SK 0763
Darlaston Pool	SJ 8833	Fazeley	SK 2001
Darleyoak Farm	SK 1422	Feltysitch	SK 0359
Daw End	SK 0300	Fenton	SJ 8944
Deepdale	SK 0853	Fisherwick	SK 1709
Deep Hayes Reservoir	SJ 9653	Five Lanes Ends	SJ 8525
Delph	SO 9186	Flash	SK 0267
Denstone	SK 0940	Fole	SK 0437
Dilhorne	SJ 9743	Folly	SK 0462
Dimminsdale	SK 0543	Ford Hayes	SJ 9246

Forest Banks	SK 1228	Hanbury	SK 1727
Forton	SJ 7521	Hanchurch	SJ 8441
Foucher's Pool	SO 8488	Hanchurch Hills	SJ 8340
Four Ashes Hall	SO 8087	Hanchurch Pools	SJ 8440
Four Crosses	SJ 9509	Hanch Wood	SK 1014
Foxholes	SK 1521	Handford Bridge	SJ 8642
Fox Inn (Seisdon)	SO 8296	Hand Leasow Wood	SK 0230
Fradley	SK 1513	Handsworth	SP 0590
Fradswell	SJ 9931	Hanford	SJ 8642
Freeford Pool	SK 1307	Hanley	SJ 8847
Froghall	SK 0247	Hanyards Lane	SJ 9523
Fullmoor Wood	SJ 9411	Harborne	SP 0284
		Harborne Reservoir	SP 0383
Gailey Reservoir	SJ 9310	Hardings Booth	SK 0664
Gayton	SJ 9828	Hardiwick Heath	SJ 9432
George's Hayes	SK 0613	Hardiwick Holly Wood	SJ 9333
Gib Torr	SK 0264	Harlaston	SK 2110
Gibbet Lane	SO 8683	Harpfields	SJ 8545
Gilleon's Hall	SK 1022	Hart's Farm	SK 0923
Gipsy Bank	SK 1456	Hartshill	SJ 8645
Gnosall	SJ 8220	Hatherton	SJ 9610
Godstone	SK 0134	Hatton Pumping Station	SJ 8236
Goldsitch Moss	SK 0164	Haughton	SJ 8620
Gornal Wood	SO 9190	Haunton	SK 2310
Goscote	SK 0102	Hawkesyard	SK 0616
Gospel End	SO 9093	Hawk Hills	SK 1422
Gradbach	SJ 9965	Hawksmoor Wood	SK 0344
Gratton Hill	SK 1357	Hay Head Farm	SP 0498
Gratwich	SK 0231	Haywood Mill	SJ 9922
Great Barr	SP 0495	Haywood Park Farm	SJ 9920
Great Bridge	SO 9792	Heath Farm	SJ 9309
Great Bridgeford	SJ 8827	Heathylee	SK 0-6-
Greatgate	SK 0540	Heaton	SJ 9562
Greatgate Wood	SK 0440	Hednesford	SK 0012
Great Haywood	SJ 9922	Heleigh Castle	SJ 7746
Greatwood	SJ 7830	Henhurst Wood	SK 2124
Greaves Wood	SK 1527	Hextons Farm	SO 7582
Grindon	SK 0854	Hey Sprink	SJ 7842
Grindon Moor	SK 0655	Highfield	SK 0739
Groundslow Fields	SJ 8637	Highgate Common	SO 8389
Gun	SJ 9761	High Hall	SJ 8111
		High Lanes Farm	SJ 8032
Hadley End	SK 1320	High Offley	SJ 7826
Halfhead	SJ 8729	High Onn Wood	SJ 8116
Hall Dale	SK 1353	Highshutt	SK 0343
Hamps Valley	SK 0-5-	Hill Chorlton	SJ 8039
Hamstall Ridware	SK 1019	Hillend	SO 8095
Hamstead	SP 0592	Hill Ridware	SK 0817

Hill Top	SO 9993	Kinver	SO 8483
Himley	SO 8791	Kinver Bridge	SO 8483
Himley Wood	SO 8991	Kinver Edge	SO 8383
Hinkley Wood	SK 1250	Knightley Gorse	SJ 8125
Hints	SK 1502	Knightley Park	SK 1923
Hoar Cross	SK 1323	Knighton Reservoir	SJ 7328
Hobb Lane	SK 0829	Knutton	SJ 8346
Hollies Common	SJ 8222	Knypersley	SJ 8856
Hollington	SK 0538	Knypersley Pool	SJ 8955
Hollinsclough	SK 0666		
Holly Bank Cottage	SK 1619	Langley (Lower Penn)	SO 8696
Hollybank Lane	SK 0616	Langley Lawn	SJ 8406
Holly Bush Park	SK 1326	Lapley	SJ 8712
Holm Cottage	SK 0642	Leek	SJ 9856
Hoo Mill	SJ 9924	Leekbrook	SJ 9853
Hopton	SJ 9426	Leekfrith	SJ 9-6-
Hopton Pools	SJ 9525	Leigh	SK 0135
Hopwas Wood	SK 1705	Leycett	SJ 7946
Horninglow	SK 2425	Lichfield	SK 1109
Horton	SJ 9457	Little Aston	SK 0800
Hovel Covert	SK 1201	Little Bridgeford	SJ 8727
Huddlesford	SK 1509	Little Gorse	SJ 9026
Huntington	SJ 9713	Little Haywood Bridge	SK 0021
Huntley	SK 0041	Little Lintus Wood	SK 1312
Hurst Farm	SK 0923	Lode Mill	SK 1455
Hurt's Wood	SK 1353	Loggerheads	SJ 7335
		Long Birch Farm	SJ 8705
Idlerocks	SJ 9337	Longdon	SK 0714
Ilam	SK 1350	Longdon Green	SK 0813
Ingestre	SJ 9724	Longnor	SK 0864
Ipstones	SK 0249	Longnor Bridge (Lapley)	SJ 8614
Ipstones Edge	SK 0450	Longsdon	SJ 9555
Iverley House Farm	SO 8781	Longton	SJ 9142
Ivetsey Bank	SJ 8310	Lord's Bridge	SK 0643
		Lower Brockhurst	SJ 8212
Jubilee Plantation	SO 7899	Lower Mill (Madeley)	SJ 7645
		Lower Mitton Farm	SJ 8815
Keele	SJ 8045	Lower Penn	SO 8696
Kennels Farm	SJ 9722	Loynton Moss	SJ 7824
Kettlebrook	SK 2103	Ludburn Ford	SK 0962
Kiddemore Green	SJ 8508	Ludchurch	SJ 9865
Kidsgrove	SJ 8454	Lush Pool (Lucepool)	SK 1519
King's Bromley	SK 1216	Lynn	SK 0704
Kingsley	SK 0146		
Kingston	SK 0629	Madeley	SJ 7744
Kingston Pool	SJ 9423	Madeley Manor	SJ 7742
Kingswinford	SO 8888	Maer	SJ 7938
Kingswood Common	SJ 8302	Maer Heath	SJ 7739

Maer Hills	SJ 7739	Newbold	SK 2119
Manifold Valley	SK 0–5–	Newborough	SK 1325
Mansty Wood	SJ 9512	Newcastle	SJ 8445
Marchington	SK 1330	Newcastle Lane	SJ 8644
Marchington Cliff	SK 1329	Newchurch	SK 1423
Marlpit Lane	SK 1143	New Inn Lane	SJ 8741
Marquis's Drive]	SK 0115	Newlands	SK 0930
Mavesyn Ridware	SK 0816	Newport	SJ 7419
Mayfield	SK 1545	Newton Road Station	SP 0293
Meaford	SJ 8835	Norbury	SJ 7823
Meaford Farm	SJ 8936	Norbury Big Moss	SJ 7824
Meerbrook	SJ 9860	North Longdon	SK 0614
Mickle Hills	SK 0908	North Street (Stoke)	SJ 8646
Middleton Green	SJ 9935	Northwood	SJ 8542
Milford	SJ 9621	Northwood Lane	SJ 8542
Milford Station	SJ 9721	Norton Bog	SK 0309
Military Cemetery	SJ 9815	Norton Canes	SK 0107
Mill Dale	SK 1354	Norton in the Moors	SJ 8951
Mill Farm (Essington)	SJ 9403	Nowall	SJ 8539
Mill Farm (Tamworth)	SK 1904		
Mill Green	SK 0823	Oakamoor	SK 0544
Mill Haft	SJ 7922	Oakedge Park	SK 0019
Mill Lane	SK 2111	Oaken	SJ 8602
Milwich	SJ 9732	Ocker Hill	SO 9793
Mitton	SJ 8815	Offleybrook	SJ 7830
Mitton Manor	SJ 8814	Offley Hay	SJ 7929
Mixon	SK 0457	Okeover Hall	SK 1548
Moddershall	SJ 9236	Okeover Park	SK 1547
Moisty Lane	SK 1131	Oldacre Valley	SJ 9718
Mons Hill	SO 9392	Oldbury	SO 9989
Moreton Grange	SK 0222	Oliver Hill	SK 0267
Moreton Moors	SJ 7817	Onecote	SK 0455
Morridge	SK 0257	Onneley	SJ 7543
Morridge Side	SK 0254	Orgreave	SK 1415
Moseley	SJ 9303	Orton	SO 8695
Moss Carr	SK 0765	Oulton	SJ 7822
Moss Lane	SK 0242	Oulton Coppice	SJ 7922
Motty Meadows	SJ 8313	Oulton Heath (Stone)	SJ 9036
Mow Cop	SJ 8557	Ousal Dale	SK 0543
Mucklestone	SJ 7237	Ouseley Cross	SK 1244
Mud-dale Wood	SK 0340	Outlanes	SJ 9035
Musden Grange	SK 1251	Outwood Hills	SK 2324
Musden Wood	SK 1151	Outwoods	SK 2225
Needwood Forest	SK 1–2–	Paradise	SJ 9206
Needwood House	SK 1825	Paradise Walk	SK 1250
Nelson Hall	SJ 8334	Parkfield	SO 9296
Nethertown	SK 1017	Pasturefields	SJ 9924

Patshull	SJ 8000	Rocester	SK 1039
Patshull Pool	SO 8099	Roches	SK 0062
Pattingham	SO 8299	Rock Lane	SJ 7336
Peasland Rocks	SK 1456	Rolleston	SK 2327
Pelsall	SK 0103	Rough Hills	SO 9296
Pendeford	SJ 8903	Rough Park Wood	SK 1219
Penkhull	SJ 8644	Rough Wood	SJ 9800
Penkridge	SJ 9214	Rowley Regis	SO 9687
Penkridge Bank	SK 0016	Royal Cottage	SK 0263
Penn Common	SO 8994	Rudyard	SJ 9557
Pensnett Reservoir	SO 9188	Rudyard Lake	SJ 9460
Perry Barr	SP 0791	Rugeley	SK 0418
Perry Hall	SP 0691	Rushall Castle	SP 0299
Perton	SO 8598	Rushley Wood	SK 1151
Perton Pool	SO 8596	Rushton	SJ 9362
Pickwood Dene (Leek)	SJ 9855		
Piggenhole	SK 0862	St Thomas	SJ 9422
Pipe Ridware	SK 0917	Sandon	SJ 9429
Podmore Pool	SJ 7735	Sandon Bank	SJ 9428
Pool Hall	SO 8597	Sandon Wood	SJ 9629
Pool Hall Pool	SO 8596	Scotch Hill	SK 1622
Pool House	SJ 8833	Seabridge	SJ 8343
Portway Lane	SK 2109	Seckley Wood	SO 7678
Pottal Reservoir	SJ 9714	Sedgley	SO 9193
Pound Green	SO 7578	Sedgley Beacon	SO 9194
Prestwood Farm	SJ 9401	Seedcroft	SK 0723
Prestwood House	SO 8686	Seighford	SJ 8725
		Seisdon	SO 8394
Quarnford	SK 0-6-	Seven Acre Wood	SK 0140
Quintin's Orchard	SK 0818	Seven Springs	SK 0020
Quixhill	SK 1041	Sharpcliffe Hall	SK 0052
		Shaw Hall	SK 0046
Raddlepits	SK 1046	Shaw Wood	SK 0031
Ramshaw Rocks	SK 0262	Shebdon	SJ 7625
Ramshorn (Ramsor)	SK 0845	Shelmore Wood	SJ 8021
Rangemore	SK 1822	Shelton	SJ 8747
Ranger	SK 0543	Shenstone	SK 1104
Ranton	SJ 8524	Sherbrook Valley	SJ 9818
Ranton Abbey	SJ 8324	Sheriff Hales	SJ 7612
Ravensclough Wood	SJ 9163	Shirleywich	SJ 9825
Ravens Tor	SK 1453	Shobnall	SK 2223
Reaps Moor	SK 0861	Short Heath	SJ 9700
Red Cow Inn	SK 0730	Shoul's Wood	SK 0628
Red Hill (Stone Park)	SJ 9134	Shugborough	SJ 9922
Restlars (Wrostlers)	SJ 8213	Shutlanehead	SJ 8242
Rickerscote	SJ 9320	Sideway (Stoke)	SJ 8743
Ridge Hill	SJ 7845	Silverdale	SJ 8246
Rileyhill	SK 1115	Sinai Park	SK 2223

Slitting Mill	SK 0217	Swithamley	SJ 9764
Smallwood Manor	SK 1029	Swynnerton	SJ 8535
Smestow	SO 8591	Sycamores Hill	SJ 9718
Smestow Gate	SO 8492	Syerscote Manor	SK 2207
Snowdon Pool	SJ 7801		
Soles Hill	SK 0952	Tamebridge	SP 0195
Solomon's Hollow	SK 0058	Tamhorn Park Farm	SK 1807
Sparrowlee	SK 0951	Tamworth	SK 2004
Spring Fields	SJ 8544	Tamworth Castle	SK 2003
Springfield Tileries	SJ 8644	Tatenhill	SK 2022
Springpool Wood	SJ 8243	Tean	SK 0039
Spring Valley	SJ 8639	Teddesley Hay	SJ 9-1-
Stafford	SJ 9223	Teddesley Park	SJ 9415
Stafford Castle	SJ 9022	Tettenhall	SO 8799
Standcliff Farm	SK 1545	Thatchmore	SK 1521
Standon	SJ 8134	The Dearndales	SK 0732
Standon Hall	SJ 8035	The Folly	SJ 7235
Stanley Pool	SJ 9351	The Hattons	SJ 8804
Stanton	SK 1246	The Hollies	SO 8085
Stanton Wood	SK 1345	The Radfords	SJ 9034
Stapenhill	SK 2521	The Stretters	SJ 8536
Star Inn	SK 0645	The Wergs	SJ 8700
Star Wood	SK 0545	Thorpe Constantine	SK 2508
Statfold Hall	SK 2307	Thor's Cave	SK 0954
Steenwood Cottages	SK 0522	Thorswood	SK 1147
Stewponey	SO 8684	Threap Wood	SK 0443
Stile Cop	SK 0315	Three Lows	SK 0746
Stockton Brook	SJ 9152	Three Shires Head	SK 0068
Stoke	SJ 8845	Throwley Moor	SK 1052
Stone	SJ 9033	Throwley Old Hall	SK 1152
Stonefield	SJ 8934	Tipton	SO 9692
Stourbridge	SO 8984	Tittensor	SJ 8738
Stowe Pool	SK 1210	Tittensor Common	
Streetly	SP 0999	(Chase)	SJ 8736
Stretton	SK 2526	Tittensor Hills	SJ 8736
Stretton Bridge	SJ 8910	Tividale	SO 9690
Stubbs Wood	SJ 8425	Tixall	SJ 9722
Sudbury	SK 1631	Tower Hill Farm	SP 0592
Sugarloaf	SK 0956	Travellers Rest	SK 0367
Sugnall Park	SJ 8030	Trentham Lake	SJ 8640
Sunny Bank	SK 1354	Trentham Park	SJ 8540
Sutton	SJ 7922	Trentham Station	SJ 8841
Swainsley	SK 0957	Trescott	SO 8497
Swainsmoor	SK 0261	Trysull	SO 8594
Swallow Moss	SK 0659	Tuppenhurst Lane	SK 1014
Swindon	SO 8690	Turbine Cottage	SO 8681
Swineholes Wood	SK 0450	Turner's Pool	
Swinfen	SK 1-0-	(Tamworth)	SK 2102

Tutbury	SK 2129	Whitmore	SJ 8040
Tutbury Castle	SK 2029	Whitmore Common	
Tyrley Wharf	SJ 6932	(Heath)	SJ 7941
		Whitmore Wood	SJ 7941
Upper Arley	SO 7680	Whittington (Kinver)	SO 8682
Upper Cadlow	SK 0450	Whittington (Lichfield)	SK 1508
Upper Cotton	SK 0547	Wigginton House	SK 2005
Upper Penn	SO 8995	Wigginton Lodge	SK 1905
Uttoxeter	SK 0833	Wightwick	SO 8798
		Willingsworth Furnaces	SO 9794
Wall	SK 0906	Willoughbridge	SJ 7439
Wall Grange	SJ 9653	Windsend	SJ 7928
Wall Heath	SO 8889	Windswell Pool	SJ 7521
Walsall	SP 0198	Winkhill	SK 0651
Walsall Wood	SK 0403	Winnoth Dale	SK 0340
Walton (Stone)	SJ 9033	Winshill	SK 2623
Waltonhurst	SJ 8527	Wolfscote Dale	SK 1357
Walton on Trent	SK 2118	Wolseley	SK 0220
Walton Station	SK 2018	Wolseley Bridge	SK 0220
Walton's Wood	SJ 7846	Wolseley Park	SK 0119
Warslow	SK 0858	Wolstanton	SJ 8548
Warwickshire Moor	SK 2104	Wolstanton Marsh	SJ 8547
Washgate	SK 0567	Wolverhampton	SO 9099
Waterhouses	SK 0850	Wombourn	SO 8793
Weag's Bridge	SK 1054	Wood Lane	SJ 9801
Weaver Hills	SK 0946	Woodmill	SK 1320
Wednesbury	SO 9894	Woodseaves	SJ 7925
Wednesfield	SJ 9400	Woodside	SO 9288
Weeford	SK 1303	Wootton	SK 1045
West Bromwich	SO 9-9-	Wootton Park	SK 0944
Weston on Trent	SJ 9727	Wordsley	SO 8987
Weston Park	SJ 8010	Wren's Nest	SO 9391
Weston under Lizard	SJ 8010	Wrinehill	SJ 7547
Wetley Rocks	SJ 9649	Wrinehill Wood	SJ 7544
Wetmore	SK 2524	Wrostler (Restlars)	SJ 8213
Wetton	SK 1055	Wrottesley	SJ 8-0-
Wetton Mill	SK 0956	Wychnor	SK 1716
Wheaton Aston	SJ 8512		
Whiston	SK 0447	Yarlet	SJ 9128
Whiston Eaves	SK 0446	Yarnfield	SJ 8632
Whitemere Bog	SK 1421	Yoxall	SK 1419
White Sitch Pool	SJ 7912	Yoxall Bridge	SK 1317
Whitgreave	SJ 8928	Yoxall Lodge	SK 1522
Whitleyford Bridge	SJ 7423		

BIBLIOGRAPHY

The following abbreviations are used:
BEC Rep: Botanical Exchange Club Report
BSBI: Botanical Society of the British Isles
JB: Journal of Botany (London)
JLS: Journal of the Linnean Society of London
NSJFS: North Staffordshire Journal of Field Studies
TNSFC: Transactions of the North Staffordshire Field Club
WBEC: Watson Botanical Exchange Club

The list below includes the most important books and articles with a bearing on the Staffordshire flora, but there are some additional items in Simpson (1960) and also the annual botany reports in TNSFC (1889–1960) and NSJFS (1961–70)

Babington, A. M. ed (1897). *Memorials Journal and Botanical Correspondence of Charles Cardale Babington*, Cambridge
Bagnall, J. E. (1895). 'New Staffordshire plants', *JB*, 33, 283
Bagnall, J. E. (1901). 'The flora of Staffordshire', *JB*, 39, supplement
Bagnall, J. E. (1908). 'Botany', *Victoria History of the County of Staffordshire* (ed W. Page), 1, 41–60
Barns, T. (1912). 'Plants found at Hilderstone', *TNSFC*, 46, 117
Berrisford, S. and Walker, A. (1906). 'Plants found on the railway embankment, Oakamoor', *TNSFC*, 40, 73
Bloxam, A. (1853). 'Wild British plants in the neighbourhood of Warslow, Staffordshire', *Phytologist* (new series), 1 (1855), 75
Booker, L. (1825). *A Descriptive and Historical Account of Dudley Castle*, 107–9
Brown, E. (1863). 'The flora of the district surrounding Tutbury and Burton-on-Trent', *The Natural History of Tutbury* (O. Mosley), 231–365
Brown, N. E. (1887). '*Vaccinium intermedium* Ruthe, a new British plant', *JLS*, 24 (1889), 125–8
Burges, R. C. L. (1944). 'Adventive flora of Burton-on-Trent', *BEC 1943–4 Rep* (1946), 815–19
Burton Flora (1896–1901). 'The flora of Burton-on-Trent and neighbourhood', *Transactions of Burton-on-Trent Natural*

History and Archaeological Society, 3, 177–90, 269–82, 4, 75–88, 117–48

Carter, J. (1839). 'A few observations on some of the natural objects in the neighbourhood of Cheadle, Staffordshire', *Magazine of Natural History* (new series), 3, 72–6

Clapham, A. R. ed (1969). *Flora of Derbyshire*, Derby

Clifford, T. and A. (1817). 'Flora Tixalliana', *A Topographical and Historical Description of the Parish of Tixall in the County of Stafford*, 285–308, Paris

Curtis, R. (1930). 'Adventive flora of Burton-upon-Trent', *BEC 1930 Rep* (1931), 465–9

D (name unknown), A. M. (1866). '*Fritillaria meleagris*', *Science-Gossip*, 2 (1867), 186

Dandy, J. E. (1958). *List of British Vascular Plants*

Dandy, J. E. (1969). 'Nomenclatural changes in the *List of British Vascular Plants*', *Watsonia*, 7, 157–78

Derham, Wm. (1760). *Select Remains of the learned John Ray*

Dickenson, S. (1798). 'A catalogue of plants ascertained to be indigenous in the county of Stafford', *The History and Antiquities of Staffordshire* (S. Shaw), 1, 97–115

Druce, G. C. (1932). *The Comital Flora of the British Isles*, Arbroath

Edees, E. S. (1941). 'The genus *Euphrasia* in north Staffordshire', *JB*, 79, 184–7

Edees, E. S. (1944). 'The first Staffordshire Flora', *North Western Naturalist*, 19, 143–6

Edees, E. S. (1944). 'Notes on Staffordshire plants', *North Western Naturalist*, 19, 272–7

Edees, E. S. (1948). 'The early history of field botany in Staffordshire, 1597–1839', *TNSFC*, 82, 81–110

Edees, E. S. (1950). 'Robert Garner, 1808–1890', *TNSFC*, 84, 13–45

Edees, E. S. (1955). 'Notes on Staffordshire brambles', *BSBI Proceedings*, 1, 301–11

Edees, E. S. (1961). 'Three common alien plants in Staffordshire', *NSJFS*, 1, 115–18

Edees, E. S. (1967). 'The wild orchids of Staffordshire', *NSJFS*, 7, 4–26

Edwardes, D. (1877). 'The wild flowers of north Staffordshire' (including a list of plants found near Denstone in 1876), *TNSFC*, 11, 48

Edwardes, D. (1878). 'The autumn wild flowers of north Staffordshire', *TNSFC*, 12, 77.

Edwards, E. (1876). Note on *Crocus nudiflorus* in Staffordshire, *JB*, 14, 347

Fiennes, C. (1698). *The Journeys of Celia Fiennes* (C. Morris, 1947)

Fraser, J. (1864). 'Plants found in Staffordshire, 1864', *Transactions of the Dudley and Midland Geological and Scientific Society and Field Club*, 2 (1865), 56–72

Garner, R. (1844). *The Natural History of the County of Stafford*, 333–445

Garner, R. (1855). *Eutherapeia*, 209 footnote

Garner, R. (1857). *Trentham and its Gardens* (Anon), botanical information attributed to Garner

Garner, R. (1871). 'North Staffordshire tracts', *Staffordshire Advertiser*

Garner, R. (1872). 'A curious British plant', *Science-Gossip*, 8 (1873), 248–9

Garner, R. (1878). 'The Staffordshire flora', *TNSFC*, 12, 95

Gerarde, J. (1597). *The Herball or Generall Historie of Plantes*

Gibson, E. (1695). *Camden's Britannia*, 539

Gisborne, J. (1797). *The Vales of Wever*

Gisborne, T. (1794). *Walks in a Forest*

Goodall, R. W. (1882). 'The wild flowers of north Staffordshire', *TNSFC*, 16, 77

Gourlay, W. B. (1919). '*Vaccinium intermedium* Ruthe', *JB*, 57, 322

Gourlay, W. B. and Vevers, G. M. (1919). '*Vaccinium intermedium* Ruthe', *JB*, 57, 259–60

Grigson, G. (1958). *The Englishman's Flora*

Hall, F. T. and R. H. (1940). 'Notes on the flora of Buxton and district', *BEC 1939–40 Rep* (1942), 338–55

Ick, W. (1836). 'Remarkable plants found growing in the vicinity of Birmingham in the year 1836', *Analyst*, 6 (1837), 20–8

Jackson, M. A. (1837). 'Catalogue of some of the rarer species of plants found in the neighbourhood of Lichfield', *Analyst*, 6, 297–8

Johnson, T. (1641). *Mercurius Botanicus*

Masefield, J. R. B. (1909). 'Staffordshire ferns', *TNSFC*, 43, 102

Mathews, W. (1884). 'List of midland county plants 1849–1884'. *Transactions of the Worcestershire Naturalists' Club* (1897–9), 55–102

Meikle, R. D. (1952). '*Salix calodendron* Wimm. in Britain', *Watsonia*, 2, 243–8

258 BIBLIOGRAPHY

Moore, C. (1897). 'A list of plants seen in the neighbourhood of Stafford, King's Bromley etc', *TNSFC*, 31, 74

Myers, J. (1945). 'Staffordshire', *The Land of Britain*, part 61, 569–652

Newman, E. (1843). 'County lists of the British ferns and their allies', *Phytologist*, 1, 508–9

Newton, A. (1971). 'Six brambles (*Rubi*) from the north midlands', *Watsonia*, 8, 369–77

Nowers, J. E. and Wells, J. G. (1890). 'Notes on a salt marsh at Branston', *Transactions of Burton-on-Trent Natural History and Archaeological Society*, 2, 50–7

Observator (1786). 'Topographical description of Shareshill', *Gentleman's Magazine*, 56, 409

Painter, W. H. (1889). *The Flora of Derbyshire*

Painter, W. H. (1892). 'The botany of Biddulph', *Midland Naturalist*, 15, 131–9, 162–4, 183–9

Painter, W. H. (1897). 'A list of plants seen within five miles of Biddulph church, 1885–92', *TNSFC*, 31, 74

Perring, F. H. and Walters, S. M. eds (1962). *Atlas of the British Flora*

Pigott, C. D. (1969). 'The status of *Tilia cordata* and *T. platyphyllos* on the Derbyshire limestone', *Journal of Ecology*, 57 no 2, 491–504

Pitt, Wm. (1794, 1796). *General View of the Agriculture of the County of Stafford*, preliminary report and first edition

Pitt, Wm. (1817). *A Topographical History of Staffordshire*, Newcastle-under-Lyme

Plot, R. (1686). *The Natural History of Staffordshire*, Oxford

Pratt, A. (1899). *The Flowering Plants of Great Britain*

Purchas, W. H. (1879). 'On *Symphytum asperrimum*', *JB*, 17, 85

Purchas, W. H. (1885). 'Some more notes on Dovedale plants', *JB*, 23, 181–4, 196–203

Purchas, W. H. (1899). '*Hieracium cymbifolium* sp. n.', *JB*, 37, 421

Purchas, W. H. and Ley, A. (1889). *A Flora of Herefordshire*, Hereford

Purton, T. (1817). *A Midland Flora*, Stratford-upon-Avon

Purton, T. (1821). *An Appendix to the Midland Flora*

Ray, J. (1670). *Catalogus Plantarum Angliae et Insularum Adjacentium*

Ray, J. (1724). *Synopsis Methodica Stirpium Britannicarum*, 3rd edition, edit J. Dillenius

Reader, H. P. (1922). Appendix to botany report on some rare
Staffordshire plants, *TNSFC*, 56, 111
Reader, H. P. (1923). 'The flora of Hawkesyard', *TNSFC*, 57, 105–17
Riddelsdell, H. J. and Baker, E. G. (1906). 'British forms of *Heli-osciadium nodiflorum* Koch', *JB*, 44, 187
Ridge, W. T. B. (1922–9). 'The flora of north Staffordshire', *TNSFC*,
56–63, appendices 1–8
Scott, Wm. (1832). *Stourbridge and its Vicinity*, 539–58
Shaw, S. (1801). 'Additions and corrections to the general history',
The History and Antiquities of Staffordshire, 2
Shimwell, D. W. (1968). 'The vegetation of the Derbyshire dales',
Nature Conservancy report
Simpson, N. D. (1960). *A Bibliographical Index of the British Flora*,
225–8, Bournemouth
Turner, D. and Dillwyn, L. W. (1805). *The Botanist's Guide through
England and Wales*, 2, 532–5
Tylecote, E. T. (1886). 'Chartley and Stowe', *TNSFC*, 20, 35–41
Tylecote, J. H. (1885). 'Plants of Stone, Yarlet, Weston and Sandon',
TNSFC, 19, 26
Waring, R. (1770). 'A letter . . . to Daines Barrington on some plants
found in several parts of England', *Philosophical Transactions*,
61 (1772), 359–89
Watson, H. C. (1835, 1837). *The New Botanist's Guide*, 1 (1835),
207–9, 2 (1837), 623–4
Watson, H. C. (1847–59). *Cybele Britannica*
Watson, H. C. (1873–4, 1883). *Topographical Botany*, eds 1–2
Watson, W. C. R. (1958). *Handbook of the Rubi of Great Britain and
Ireland*, Cambridge
Wilmott, A. J. (1942). '*Phyteuma spicatum* L. in Staffordshire', *JB*,
80 (1944 for 1942), 135
Withering, Wm. (1776, 1787, 1796, 1801). *An Arrangement of British
Plants*, eds 1–4 (with variations of title), Birmingham
Withering, Wm. junior (1822). *The Miscellaneous Tracts of the late
William Withering*
Wolley-Dod, A. H. (1931). 'A revision of the British roses', *JB*, 68–9
(1930–1), supplement
Wright, M. (1934). *The Best of Cannock Chase*
Yeo, P. F. (1970). '*Euphrasia brevipila* and *E. borealis* in the British
Isles', *Watsonia*, 8, 41–4

MANUSCRIPTS

Berrisford MS. Records by S. Berrisford of Oakamoor in a copy of Anne Pratt's *The Flowering Plants of Great Britain* (1899–1900), which passed into the possession of his daughter, Mrs J. Plant.

Bloxam MS. Records by A. Bloxam in a copy of O. Mosley's *The Natural History of Tutbury* (1863), which now belongs to R. G. Warren.

Carrington Drawings. A notebook of pen drawings and water colour sketches of wild flowers from the vicinity of Wetton by S. Carrington (undated but probably about 1835) in the present author's library.

Forster 1796. 'A list of some of the more rare indigenous plants of the county of Stafford' by R. Forster of Stone, dated 14 October 1796, in the William Salt Library, Stafford.

Garner MS. R. Garner's own interleaved copy of *The Natural History of the County of Stafford* (1844) in the William Salt Library.

Gisborne MS. Records by T. Gisborne of Yoxall Lodge in a copy of W. Hudson's *Flora Anglica* (1778) in the library of Cambridge University.

Levett MS. Notes by Miss L. F. Levett (undated but probably about 1850) in the possession of Mrs D. R. Haszard of Milford Hall, Staffordshire.

Power MS. Records by J. A. Power in a copy of Turner and Dillwyn's *The Botanist's Guide* (1805) in the library of the National Museum of Wales.

Ridge MS. W. T. B. Ridge's bound and interleaved copy of 'The flora of north Staffordshire' (1922–9) in the present author's possession.

Smith 1871. A hand-made notebook inscribed, 'Catalogue of the more rare and remarkable flowering plants and ferns of Dovedale and its neighbourhood, by G. Smith, Ockbrook, 26 May 1871', owned by Miss K. M. Hollick.

In addition there are lists of plants in the library at Kew which were sent to H. C. Watson for incorporation into *Topographical Botany*.

The author's personal notes, which include hundreds of localised lists as well as record cards and distribution maps, are kept in his own house.

HERBARIA

Staffordshire specimens are to be found here and there in many of the large public and university herbaria and no doubt in several private herbaria too. The following is a list of the local collections referred to in this book.

Bagnall, J. E. (1830–1918) Birmingham City Museum and Art Gallery. Includes a few specimens collected by J. Fraser. Specimens collected by A. Harlond have been added to the original collection.

Berrisford, S. (1859–1938) Incorporated in the author's herbarium.

Daltry, H. W. (1887–1962) Incorporated in the author's herbarium. Includes specimens collected by J. Daltry and T. W. Daltry.

Edees, E. S. The author's herbarium which is kept at his home address.

Fraser, J. (1820–1909) University of Hull.

Gisborne, T. (1758–1846) British Museum (Natural History).

Masefield, J. B. R. (1850–1932) Incorporated in the author's herbarium.

Moore, C. (1870–1944) Mid-Staffordshire Field Club.

Nowers, J. E. (1864–1947) Darlington and Teesdale Naturalists' Field Club.

Power, J. A. (1810–86) Holmesdale Natural History Club Museum, Reigate.

Reader, H. P. (1850–1929) Stoke on Trent City Museum and Art Gallery. Includes plants collected by P. P. Thornton, L. K. Clark and F. D. Murray. Others collected by P. Allen have been added to the herbarium.

Ridge, W. T. B. (1872–1943) Incorporated in the author's herbarium.

ACKNOWLEDGEMENTS

It is a very pleasant duty at the end of a long task to acknowledge the help of many friends and correspondents. The author is indebted to all who have sent him records and lists of plants. Their names are included in the list of recorders (p 23). Special thanks are due to G. J. V. Bemrose, B. Bentley, H. W. Daltry, W. H. Hardaker and V. Jacobs for help in the early days, to Miss K. M. Hollick and the brothers G. A. and M. A. Arnold for contributions which continue to the present day and to R. H. Brown, not only for his remarkable discoveries, but also for the infection of his enthusiasm.

Throughout the years, many botanists with specialist knowledge have willingly named difficult plants and the author would here record his thanks to the following for help during the preparation of this Flora: P. M. Benoit (*Callitriche*), Dr C. D. K. Cook (*Ranunculus* subgen *Batrachium*), J. A. Crabbe (*Polypodium*), M. G. Daker (*Fumaria*), J. E. Dandy (*Potamogeton*), R. C. L. Howitt (*Salix*), Dr C. E. Hubbard (*Gramineae*), R. D. Meikle (*Salix*), R. Melville (*Ulmus*), Dr F. H. Perring (*Arctium*), Dr A. J. Richards (*Taraxacum*), P. D. Sell (*Hieracium*), Mrs H. R. H. Vaughan (*Rosa*), A. E. Wade (*Symphytum*), Dr C. West (*Hieracium*), Dr P. F. Yeo (*Euphrasia*). Dr J. G. Dony, N. D. Simpson and Dr D. W. Shimwell have also given valuable help in different ways.

The author is very grateful to Professor S. H. Beaver of Keele University and G. Barber, the chief technician of the Geography Department, for the five maps which illustrate the introduction to his book, and to J. T. Ellis, G. B. Burgess, R. H. Hall and the Leek and Westbourne Building Society for photographs.

He would also like to express his thanks to the publishers and their botanical editor for their help and courtesy during the preparation of his work for the press and to Dr S. M. Walters for reading the proofs.

Last of all and above all there is the companion of all his travels, whose encouragement has never flagged and whose name ought to appear with his own as co-author of this book.

INDEX
of the genera and English names

CORRECTION

It is regretted that there are a few minor discrepancies between some of the distribution maps and the statistical summaries in the text, due to the accidental omission of a circle or the filling in of one which should have been left hollow. The differences are not important, but where they occur the statement in the text is authoritative.